Springer Theses

Recognizing Outstanding Ph.D. Research

For further volumes:
http://www.springer.com/series/8790

Aims and Scope

The series "Springer Theses" brings together a selection of the very best Ph.D. theses from around the world and across the physical sciences. Nominated and endorsed by two recognized specialists, each published volume has been selected for its scientific excellence and the high impact of its contents for the pertinent field of research. For greater accessibility to non-specialists, the published versions include an extended introduction, as well as a foreword by the student's supervisor explaining the special relevance of the work for the field. As a whole, the series will provide a valuable resource both for newcomers to the research fields described, and for other scientists seeking detailed background information on special questions. Finally, it provides an accredited documentation of the valuable contributions made by today's younger generation of scientists.

Theses are accepted into the series by invited nomination only and must fulfill all of the following criteria

- They must be written in good English.
- The topic should fall within the confines of Chemistry, Physics, Earth Sciences, Engineering and related interdisciplinary fields such as Materials, Nanoscience, Chemical Engineering, Complex Systems and Biophysics.
- The work reported in the thesis must represent a significant scientific advance.
- If the thesis includes previously published material, permission to reproduce this must be gained from the respective copyright holder.
- They must have been examined and passed during the 12 months prior to nomination.
- Each thesis should include a foreword by the supervisor outlining the significance of its content.
- The theses should have a clearly defined structure including an introduction accessible to scientists not expert in that particular field.

Maik Rudolf Johann Scherer

Double-Gyroid-Structured Functional Materials

Synthesis and Applications

Doctoral Thesis accepted by
University of Cambridge, UK

Springer

Author
Dr. Maik Rudolf Johann Scherer
Cavendish Laboratory
Department of Physics
University of Cambridge
Cambridge
UK

Supervisor
Prof. Ullrich Steiner
Cavendish Laboratory
Department of Physics
Univeristy of Cambridge
Cambridge
UK

This work was funded through the Nokia-Cambridge University Strategic Research Alliance in Nanotechnology.

ISSN 2190-5053 ISSN 2190-5061 (electronic)
ISBN 978-3-319-00353-5 ISBN 978-3-319-00354-2 (eBook)
DOI 10.1007/978-3-319-00354-2
Springer Cham Heidelberg New York Dordrecht London

Library of Congress Control Number: 2013937194

Printed on acid-free paper

Springer is part of Springer Science+Business Media (www.springer.com)

Insanity: doing the same thing over and over again and obtaining different results.[1]

[1] Original quotation by Albert Einstein: "Insanity: doing the same thing over and over again and expecting different results." North American Wildlife and Natural Resources Conference, Volume 71, p. 54, Wildlife Management Institute, 1975.

Dedicated to my family

Supervisor's Foreword

The detailed control of material assembly on nanometer length scales is an important strategy in the optimization of energy-related technologies. Electronic processes in organic and organic–inorganic hybrid solar cells, in batteries and in super-capacitors take place at interfaces. An optimal device architecture therefore has to simultaneously fulfil two requirements: the maximization of interfacial area per volume, and the direct connectivity of each interfacial element with the outside circuit. This is a considerable scientific and technological challenge since it requires the local control over materials assembly on the 10 nm length scale in layers that are many micrometers or even millimeters thick.

Maik Scherer's doctoral theses explore the use of polymer self-assembly for the manufacture of energy materials with well-defined nano-morphologies and structural connectivities. To this end, the self-assembly of block-copolymers into the double-gyroid morphology was optimized for use as a 10-nm scaffold for material synthesis. This addresses the challenge in the field of polymer self-assembly of using these highly periodic morphologies in the synthesis of functional materials. It is demonstrated in a range of different ways. Metals and metal-oxides were introduced into voided gyroid channel networks by electroplating from a substrate or by atomic layer deposition; single crystals were grown within the mesoporous network; and conjugated polymers were electropolymerized within the gyroid void channels.

This opened-up a number of strategies for the creation of high-performance functional materials whose morphology is controlled on the 10-nm length scale, which Maik used to create materials with unprecedented electrochromic properties, a world-leading supercapacitor, and 3D 10 nm structured conjugated polymer assemblies that cannot be manufactured in any other fashion.

Finally the strategies developed in this doctoral thesis present a generic route to the synthesis of a very wide array of materials—both inorganic and organic—with periodic 3D morphologies on the 10-nm length scale, allowing the structural optimization of a large range of functional materials.

Cavendish Laboratory, Cambridge April 2013 Prof. U. Steiner

Preface

Inspired by the work of Edward Crossland et al. on using a thin porous polymer film with double-gyroid morphology for the template-assisted precision pattering of inorganic materials on the 10 nm length scale, [1–5] and due to the fact that all of this polymeric material which was custom synthesized in small quantities by Prof. Dr. Marc Hillmyer had been used up, I established a recipe for preparing thin gyroid films using commercially available polymers during my graduate studies [6, 7]. However, the latter system had some shortcomings as will be discussed below, thus the supply of double gyroid templates for the patterning of functional materials remained a challenge.

Neat diblock copolymers form the desired double-gyroid morphology during microphase separation only for a very narrow range of block volume ratios. Nevertheless, Prof. Hillmyer succeeded in synthesizing a pure diblock copolymer, poly(4-fluorostyrene)-*b*-poly(D,L-lactide) (PFS-*b*-PLA), which adopted the gyroid morphology in thin films during temperature annealing [1–5]. In contrast, the system I investigated was a binary blend of a lamellae forming diblock copolymer, poly(styrene)-*b*-poly(isoprene) (PS-*b*-PI), and a homopolymer, poly(styrene) (hPS), which forms a gyroidal film during slow solvent-casting from toluene [6, 7]. The solvent evaporation and thereby the film formation is controlled by sandwiching the polymer solution between a teflon foil and the substrate.

The main advantage of the PS-*b*-PI/hPS system is that commercially available copolymers of a wide block composition range can be used since the final blend composition, which determines the microphase separated morphology, can be altered by addition of hPS homopolymer. However, compared to the neat PFS-*b*-PLA the PS-*b*-PI/hPS blend has some major drawbacks. The preparation of templates using PS-*b*-PI/hPS is very time-consuming and involves a drying process that typically takes several days, while the PFS-*b*-PLA films only require brief temperature annealing for less than an hour. Further, film areas of homogeneous thickness are limited to a square centimeter, while PFS-*b*-PLA templates of arbitrary dimensions can be easily fabricated by spin or blade-coating methods. But most importantly, in contrast to PFS-*b*-PLA, the blend system formed non-porous wetting layers at the air/polymer and polymer/substrate interfaces. Especially, the latter layer prevents the usage of the films as templates for the electrodeposition of functional materials, which requires the voided network to be continuous all the way down to the conducting substrate.

The objective of this work, concerning the template-assisted patterning of functional materials on the nanoscale to enhance specific material properties, is fivefold. First, to develop a highly reproducible synthesis pathway for large quantities of double-gyroid-forming diblock copolymers. Second, to replace the previously used fluorinated monomers with an inexpensive and environmentally sustainable alternative making large-scale applications feasible. Third, to establish a reliable and fast fabrication technique for fully porous gyroidal templates. Fourth, to replicate these templates with inorganic as well as organic functional materials using electrochemical and atomic layer deposition. Finally, to apply and characterize these novel double-gyroid structured materials in devices such as photovoltaic devices, electrochromic displays, and supercapacitors.

The first part of this thesis introduces the different techniques of modern nanotechnology employed in three-dimensional nanopatterning of functional materials and seeks to highlight the possibilities and advantages of using self-assembled, mesoporous polymeric templates for this purpose. Chapter 2 explores the fascinating gyroidal intermaterial dividing surfaces formed by self-assembling systems, such as microphase separating diblock copolymers. Chapter 3 focuses on the synthesis of double-gyroid-forming diblock copolymers consisting of poly (4-X-styrene) and poly(lactic acid) via atom transfer radical and organocatalytic ring-opening polymerization using a bifunctional initiator. The preparation of mesoporous templates with double-gyroid morphology based on a rapid thermal annealing protocol is described in Chap. 4. Importantly, it is shown that copolymers prepared from inexpensive, non-substituted styrene can yield fully porous thin films.

Chapter 5 reports the template-assisted electrodeposition of metal oxides, such as V_2O_5, using the previously prepared voided styrenic templates. Further, the performance of these materials as electrodes in electrochromic displays and supercapacitors is tested. Chapter 6 describes an alternative approach to obtain double-gyroid-structured metal ceramics via the post-nanostructuring thermal oxidation of electroplated metals. Nickel is chosen to demonstrate this highly versatile fabrication route and to study the nanoscale Kirkendall effect of three-dimensional structures. Further, the electrochromic performance of NiO is investigated. Chapter 7 deals with yet another electrodeposition technique, namely the electropolymerization of conjugated polymers. Identification of suitable solvents for the templated electrosynthesis and the subsequent template dissolution enable the three-dimensional nanopatterning of PPy, PEDOT, and PT. The refilling of the organic template via atomic layer deposition, including the necessary chemically modification of the styrenic scaffold for an improved thermal resistance and surface functionality, is discussed in Chap. 8.

This thesis is the result of work carried out between May 2009 and August 2012 with the Thin Films and Interfaces Group at the Cavendish Laboratory, Cambridge, UK, and was only made possible by generous funding through Nokia and with the kind help and support I received from numerous people.

First of all I would like to thank Prof. Dr. Ullrich Steiner who gave me the unique opportunity to join his group and to work in a superb scientific environment. I am deeply grateful for the constant help, support, and excellent supervision. It was a very inspiring and exciting time in Cambridge and great fun participating in the three winter schools in Austria and Switzerland.

Many thanks to Dr. Piers Andrew, Dr. Chris Bower, Dr. Di Wei, and Dr. Richard White from the Nokia Research Center, Cambridge, UK, for all the valuable discussions and fruitful collaborations.

Especially, I would like to thank Dr. Oren Scherman for the support while synthesizing the polymeric materials in his facilities at the Melville Laboratory. A further special thanks to Prof. Sir Mark Welland for granting me access to the Nanoscience Centre's laboratories and office, as well as for the invitation to the winter school in France and summer school in Japan.

Thanks to all former and current members of the Thin Films and Interfaces Group for the help, friendship, company, and great working atmosphere in the office and lab: Mathias Kolle, Katherine Thomas, Pola Goldberg-Oppenheimer, Stefan Guldin, Rosa Poetes, Nataliya Yufa, Yuan Zhou, Silvia Vignolini, Gen Kamita, Sarah Rong, Stefano Salvatore, and Xiaoyuan Sheng. Special thanks to Pedro Cunha, Alexandre Nicolas, Harry Beeson, Micheal Price, Raphael Dehmel, and Li Li for the excellent collaborations. Another thanks to my friends Pedro for proofreading and Sven Hüttner, Peter Kohn, and Kai Scherer for measuring SAXS.

I would also like to express my gratitude to the helpful support staff of the Biological and Soft Systems Sector and the Cambridge Nanoscience Centre, especially Tom Mitchell, Sue Gymer, Sue Murkett, Ian Ganney, Suresh Mistry, Pete Bone, and Owen Dunn. Thanks to Frank Biedermann and Einat Elmalem not only for being good friends, but also for the help and advice in numerous chemistry-related questions throughout the study. I would like to say thank you to Pia Sharma and my Cambridge friends for the unforgettable time we spent together in the UK!.

Finally, I want to express my gratitude to my family and my close friends for their support and for being there for me all these years. Johann Ernst, Marion Ernst, Jeanette Scherer, Kai Scherer, Nicole Scherer, and Pia Sharma without your love and tremendous support, I would have capitulated a long time ago.

August 2012 Maik R. J. Scherer

References

1. E.J.W. Crossland, S. Ludwigs, M.A. Hillmyer, U. Steiner, Soft Matter **3**(1), 94–98 (2007)
2. E.J.W. Crossland, *Block Copolymer Patterning of Functional Materials*. University of Cambridge, Cambridge, (2008)

3. E.J.W. Crossland, M. Kamperman, M. Nedelcu, C. Ducati, U. Wiesner, D.M. Smilgies, G.E.S. Toombes, M.A. Hillmyer, S. Ludwigs, U. Steiner, H.J. Snaith, Nano Lett. **9**(8), 2807–2812 (2009)
4. E.J.W. Crossland, M. Nedelcu, C. Ducati, S. Ludwigs, M.A. Hillmyer, U. Steiner, H.J. Snaith, Nano Lett. **9**(8), 2813–2819 (2009)
5. E.J.W. Crossland, S. Ludwigs, M.A. Hillmyer, U. Steiner, Soft Matter **6**(3), 670–676 (2010)
6. A.S. Finnemore, M.R.J. Scherer, R. Langford, S. Mahajan, S. Ludwigs, F.C. Meldrum, U. Steiner, Adv. Mater. **21**(38–39), 3928–3932 (2009)
7. M.R.J. Scherer, *Nanostructured Materials via Self-Assembled Templates*, University of Konstanz, Konstanz, (2009)

Contents

Abbreviations

[BMIM][TFSI]	1-Butyl-3-methylimidazolium bis(trifluoromethylsulfonyl)imide
[EMIM][BF_4]	1-Ethyl-3-methylimidazolium tetrafluoroborate
[SET3][TFSI]	Triethylsulfonium bis(trifluoromethylsulfonyl)imide
A(R)GET	Activators (re)generated by electron transfer
AFM	Atomic force microscopy
ALD	Atomic layer deposition
ATR-FTIR	Attenuated total reflection Fourier transform infrared spectroscopy
ATRP	Atom transfer radical polymerization
Bipy	2,2′-Bipyridine
CMC	Constant mean curvature
CV	Cyclic voltammetry
DBU	1,8-Diazabicycloundec-7-ene
DCM	Dichloromethane
DEZ	Diethylzinc
DG	Double-gyroid
DI water	Deionized water
EBIB	Ethyl α-bromoisobutyrate
EDLC	Electrochemical double layer capacitor
EDOT	3,4-Ethylenedioxythiophene
EDTA	Ethylenediaminetetraacetic acid
EDX	Energy-dispersive X-ray spectroscopy
FTO	Fluorine-doped tin oxide coated glass
HMDS	Bis(trimethylsilyl)amine
HMTETA	1,1,4,7,10,10-Hexamethyltriethylenetetramine
HOMO	Highest occupied molecular orbital
HPL	Hexagonally perforated lamellar
IMDS	Intermaterial dividing surfaces
IO	Inverse opal
ITO	Tin-doped indium oxide coated glass
Lactide	3,6-Dimethyl-1,4-dioxane-2,5-dione
LiTFSI	Bis(trifluoromethane)sulfonimide lithium
LUMO	Lowest unoccupied molecular orbital
Me_6TREN	Tris[2-(dimethylamino)ethyl]amine

MO	Metal oxide
MTBD	N-methylated 1,4,7-triazabicyclodecene
MTS	Methyltrichlorosilane
NKE	Nanoscale Kirkendall effect
NMR	Nuclear magnetic resonance
ODT	Order-disorder transition
OTS	Octyltrichlorosilane
P(F)S	Poly(4-fluorostyrene)-*r*-poly(styrene)
P3MT	Poly(3-methylthiophene)
PBrS	Poly(4-bromostyrene)
PBT	Poly(bithiophene)
PCBM	[6,6]-Phenyl-C_{61}-butyric acid methyl ester
PClS	Poly(4-chlorostyrene)
PDI	Polydispersity index
PEDOT	Poly(3,4-ethylenedioxythiophene)
PEN	Poly(ethylene naphthalate)
PFS	Poly(4-fluorostyrene)
PGMEA	Propylene glycol monomethyl ether acetate
PhOPh	Diphenyl ether
PI	Poly(isoprene)
PLA	Poly(D,L-lactide)
PMDETA	N,N,N′,N″,N″-pentamethyldiethylenetriamine
POZ	Primary ozonide
PPy	Poly(pyrrole)
PS	Poly(styrene)
PT	Poly(thiophene)
PTFE	Poly(tetrafluoroethylene)
PXS	Poly(4-X-styrene)
RAFT	Reversible addition fragmentation transfer
RCA	Wafer cleaning procedure developed
RIE	Reactive-ion etching
ROP	Ring-opening polymerization
RTIL	Room temperature ionic liquids
SAXS	Small angle X-ray scattering
SEC	Size exclusion chromatography
SEM	Scanning electron microscope
SG	Single-gyroid
$Sn(EH)_2$	Tin(II) 2-ethylhexanoate
SOZ	Secondary ozonide
SU-8	Epoxy-based negative photoresist
TBD	1,4,7-Triazabicyclodecene
TCM	Chloroform
TEM	Transmission electron microscopy
THF	Tetrahydrofuran
Thiourea	Bis(3,5-trifluoromethyl)phenyl cyclohexylthiourea

TIP	Titanium isopropoxide
TMA	Trimethylaluminium
TPMS	Triply periodic minimal surfaces
TRIS	2-Amino-2-hydroxymethyl-propane-1,3-diol
UV	Ultraviolet
XPS	X-ray photoelectron spectroscopy
XRD	X-ray diffraction

Chapter 1
Introduction

Broadly speaking, nanoscience refers to a field concerning the production, characterization and exploitation of materials with nanoscale dimensions in at least one dimension. Of special interest are three-dimensionally nanostructured solid-state functional materials with structural features ranging in size from 1 to 100 nm. Functional materials are generally considered as those materials which possess particular native properties required for various modern applications. These can include energy generation, energy storage, electronics, data storage, actuators, and sensors. Functional materials can be found in all material classes from polymers over metals to ceramics and composites. Structuring these materials on the nanoscale can not only enhance particular functional characteristics but also result in novel physical properties, such as quantum mechanical effects that are present neither in the bulk nor in structures with dimensions on larger length scales. These enhanced and unique material properties often arise from the ability to access both bulk and surface/interface properties.

Currently, the fabrication of well-defined nanostructures for both scientific and industrial purposes mainly relies on top-down nanolithographic processes, such as nanoimprint, X-ray, extreme ultraviolet, and electron-beam lithography [1, 2]. Most of these methods require time consuming, complex, and expensive beam sources for exposure as well as harmful corrosive chemicals for the involved etching steps. However, the resolution of the structures, the quantities of nanostructures produced, and the fabrication of complex three-dimensional structures are usually rather limited for these techniques.

Recently, bottom–up processes utilizing self-assembly have been developed that could become the next generation of nanostructure processing, as they overcome the problems listed above. In particular, self-assembling copolymers have attracted much attention as potential nanostructuring agents over the past decades and are still subject to ongoing research [3]. The most important near-term application is their use in the microelectronics industry as etch mask with pattern dimensions far below the optical diffraction limit [4].

M. R. J. Scherer, *Double-Gyroid-Structured Functional Materials*,
Springer Theses, DOI: 10.1007/978-3-319-00354-2_1,

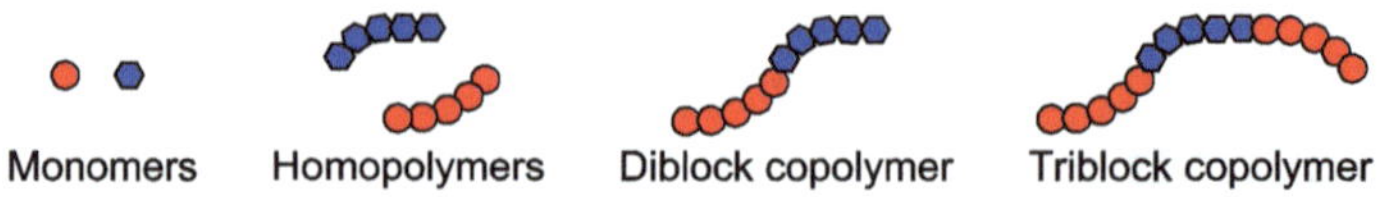

Fig. 1.1 Architectures of linear homopolymers and block copolymers

Polymers are macromolecules that consist of repeating molecular units known as monomers covalently bonded during polymerization. While various architectures are possible, linear polymers with unbranched molecular chains are the most common both naturally occurring and synthetic. Homopolymers contain only a single type of monomer, while block copolymers are made up from a sequence of chemically distinct homopolymers (Fig. 1.1).

A remarkable property of polymer systems is their ability to self-assemble, driven by thermodynamic incompatibilities of the different monomers, which will be discussed in more detail in Chap. 4. The resulting repulsive forces make homopolymer blends to separate into macrostructures. In contrast, being covalently bonded, the thermodynamically incompatible blocks of copolymers are prevented from separating on a macroscopic level. Even for the most simple copolymer architecture, namely diblocks, this gives rise to the formation of a variety of highly ordered morphologies with nanoscale periodicities [3].

The structure that is adopted at equilibrium is highly dependent on diblock copolymer composition, while the dimensions of the microdomain structure are determined by the degree of polymerization, as shown in Fig. 1.2. Symmetric diblock copolymers with equal block volumes arrange into a lamellar morphology with alternating layers of the constituent blocks. An increase of the red block's volume fraction leads to the formation of morphologies where the minority component (blue) is enclosed by the matrix phase (red). Between the lamellar and the hexagonally packed cylinders the bi-/tricontinuous double-gyroid and *Fddd* are observed, which will be discussed

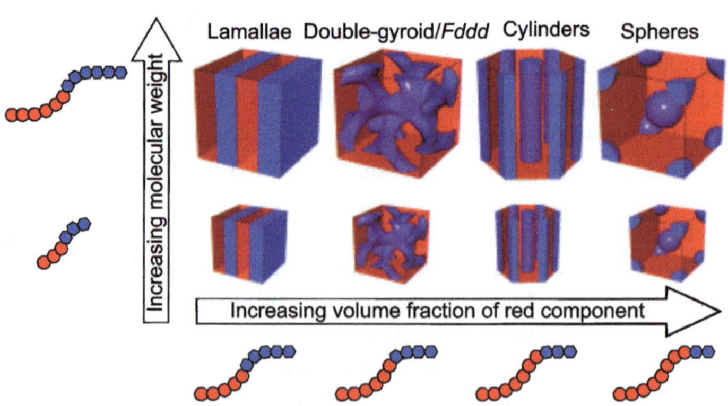

Fig. 1.2 Equilibrium phase morphologies formed by diblock copolymers. Since the phase diagram is symmetric, increasing the *blue block* results in the color-inverted structures

in Chap. 2. A further increase in asymmetry leads to the formation of body-centered cubic and close-packed spheres.

Approaches concerning the three-dimensional nanopatterning of functional materials which utilize self-assembly can be broadly divided into three categories: firstly, the direct self-assembly of the functional material; secondly, the co-self-assembly using a structure-directing agent; finally, the replication of self-assembled templates. In the following these three approaches will be compared and special attention will be paid to strategies utilizing polymer self-assembly.

By accurately controlling the synthesis and/or deposition conditions of some-functional materials, such as metals and ceramics, they can be made to independently self-assemble into nano-wires, nano-tubes, nano-forests, and nano-flowers etc., without the need of templates or structure-directing agents [5–8]. The number of materials showing this behavior is rather limited and determining the appropriate experimental conditions usually requires a protracted optimization process for each material. However, the discovery of semiconducting conjugated polymers with advanced optic and electronic properties extended this class of self-assembling functional materials. Phase separating blends of conjugated homopolymers as well as fully conjugated copolymer are extensively studied for organic solar harvesting [9, 10]. A major advantage is the possible roll-to-roll manufacturing process. Although the functional properties of the polymers can be tailored by adjusting the monomers' molecular structure, this may also alter their self-assembly characteristics [11].

An advanced nanopatterning approach of functional materials that do not independently self-assemble is to employ structure-directing agents, such as copolymers with hydrophilic and hydrophobic blocks [12]. The fact that precursors to functional materials or nanoparticles preferentially swell one of the copolymer blocks is used for co-self-assembly which is driven by the polymer's phase separation behavior. However, increasing the number of components also multiplies the complexity of the assembling system and thereby the control of structure formation. Hence, this nanopattering approach was only demonstrated for a few functional materials [13].

The most widely applicable approach splits the fabrication of nanopatterned functional materials into two successive steps, namely the self-assembly driven formation of mesoporous templates and the replication of the templates with functional materials, as illustrated in Fig. 1.3. This strategy to nanostructured functional materials has the major advantage that it allows for the independent optimization of the template preparation. Since the same template can be used to structure various functional materials applying different replication techniques, such as sol-gel, atomic layer deposition (ALD), electrochemical deposition, and electroless plating, it is considered as an extremely versatile approach [14–17].

Block copolymer self-assembly provides an elegant, cheap, and environmentally sustainable way to produce mesoporous templates. After adoption of an appropriate nanomorphology during phase separation the selective removal of one block yields the desired mesoporous template. In the past few years, a number of experimental protocols on the selective degradation of one copolymer component have been reported, see Table 1.1. Of particular interest are systems containing poly(styrene), since it is cheap, cross-linkable, chemically resistant against most bases and acids,

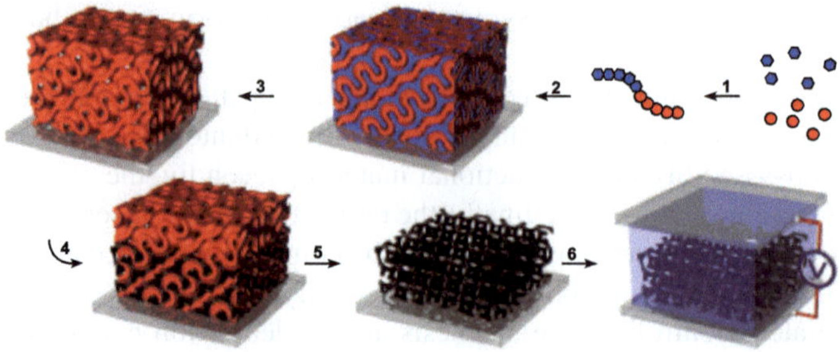

Fig. 1.3 Strategy for nanopatterning functional materials employing templates based on copolymer self-assembly with double-gyroid morphology. This approach is particularly versatile since it separates the template fabrication (*top row*) from the templating of functional materials (*bottom row*). *1* Copolymer synthesis. *2* Self-assembly into the double-gyroid morphology. *3* Selective degradation of one block yields a mesoporous template. *4* Templating of functional materials via various deposition techniques. *5* Removal of polymeric matrix. *6* Assembly of functional devices

Table 1.1 Reported techniques to selectively degrade one copolymer block. Poly(styrene) *PS*, poly(dimethylsiloxane) *PDMS*, poly(ethyleneoxide) *PEO*, poly(isoprene) *PI*, poly(lactic acid) *PLA*, poly(methyl methacrylate) *PMMA*, poly(vinylpyridine) *PVP*, 3-pentadecylphenol *PDP*

Copolymer	Degradation method	Component degraded	Reference
PS-*b*-PDMS	HF	PDMS	[18]
PS-*b*-PEO	$HI_{(aq)}$	PEO	[19]
PS-*b*-PI	Ozonolysis or UV irradiation	PI	[20, 21]
PS-*b*-PLA	$NaOH_{(aq)}$ or $HI_{(aq)}$	PLA	[22]
PS-*b*-PMMA	UV irradiation	PMMA	[23]
PS-*b*-PVP/PDP	Dissolution	PDP	[24]

and has a suitable glass transition temperature, while being easily removable after replication by combustion or dissolution in organic solvents. The glass transition temperature of about 104 °C is ideal, since it makes phase transitions at annealing temperatures below 200 °C possible while offering a large temperature processing window for the refilling of the voided styrenic matric.

When preparing mesoporous scaffolds from copolymers for the replication with functional materials it is essential that both segregated phases, which eventually will form the template matrix and the voided channels, are continuous, free-standing, and self-supporting. Thus, potentially suitable microphase structures have to be multicontinuous with the relevant multiple chemically distinct domains being fully interconnected and continuous. This ensures that neither the porous template nor its replica will collapse upon removal of the sacrificial phase or the template, respectively. For example, Crossland et al. found, that titania nanowires which were prepared by electrochemically replicating an array of voided cylinders aligned perpendicular to the substrate's surface collapse after dissolution of the polymeric template [25]. A further advantage of multicontinuous copolymer morphologies is their higher likelihood to be porous at the interfaces when casted into thin films.

The only ordered microdomain structures adopted by diblock copolymers that are multicontinuous and hence are suitable as starting material for mesoporous templates are the bicontinuous *Fddd* and tricontinuous double-gyroid. The *Fddd*'s region in the phase diagram is extremely small, as will be discussed in Chap. 4, leaving the double-gyroid as the only experimentally feasible diblock copolymer morphology [26]. Moreover, the three-dimensional double-gyroid morphology offers structural properties, such as a good structural stability, a high surface area to volume ratio, and fairly constant strut and pore diameters, which can be highly beneficial when applied in nanotechnology applications and will be discussed in Chap. 2. A first successful application of a double-gyroid-structured metal-oxide nanostructure prepared by the replication of a thin film template has been demonstrated in dye-sensitized solar cells [27]. In contrast to this earlier example, a functional thin film device based on the double-gyroid morphology that out-performs existing technologies has yet to be demonstrated.

In summary, three-dimensional nanostructuring solid-state functional materials can enhance particular functional properties important for nanotechnology applications. Self-assembly on the nanoscale can be exploited as a bottom-up approach replacing expensive lithographic nanostructuring techniques. While only a minority of the functional materials self-assemble into suitable architectures, template-assisted precision pattering utilizing self-assembled templates is extremely versatile. Polymers, especially diblock copolymers, offer a cheap and scalable route to fabricate the needed mesoporous templates. In particular, the double-gyroid seems to meet the requirements demanded of a highly performing nanostructure. This was recently discussed in a theoretical study by Khlebnikov et al. on the electronic structure of double-gyroid semiconductors for carrier multiplication solar cells [28].

References

1. B.D. Gates, Q. Xu, M. Stewart, D. Ryan, C.G. Willson, G.M. Whitesides, Chem. Rev. **105**(4), 1171–1196 (2005)
2. M. Li, C.K. Ober, Mater. Today **9**(9), 30–39 (2006)
3. R.A. Segalman, Mater. Sci. Eng **48**(6), 191–226 (2005)
4. C.T. Black, R. Ruiz, G. Breyta, J.Y. Cheng, IBM J. Res. Dev. **51**(5), 605–633 (2007)
5. L. Schmidt-Mende, J.L. MacManus-Driscoll, Mater. Today **10**(5), 40–48 (2007)
6. K.P. Musselman, G.J. Mulholland, A.P. Robinson, L. Schmidt-Mende, J.L. MacManus-Driscoll, Adv. Mater. **20**(23), 4470–4475 (2008)
7. X.-Y. Xue, Z.-H. Chen, L.-L. Xing, C.-H. Ma, Y.-J. Chen, T.-H. Wang, J. Phys, J. Phys. Chem. C **114**(43), 18607–18611 (2010)
8. S.H. Ko, D. Lee, H.W. Kang, K.H. Nam, J.Y. Yeo, S.J. Hong, C.P. Grigoropoulos, H.J. Sung, Nano Lett. **11**(2), 666–671 (2011)
9. M. Sommer, S. Hüttner, S. Wunder, M. Thelakkat, Adv. Mater. **20**(13), 2523–2527 (2008)
10. M. Sommer, S. Huettner, M. Thelakkat, J. Mater. Chem. **20**(48), 10788 (2010)
11. M. Sommer, S. Hüttner, U. Steiner, M. Thelakkat, Appl. Phys. Lett. **95**(18), 183308–183308-3 (2009)
12. M. Templin, A. Franck, A.D. Chesne, H. Leist, Y. Zhang, R. Ulrich, V. Schädler, U. Wiesner, Science **278**(5344), 1795–1798 (1997)

13. J. Lee, M.C. Orilall, S.C. Warren, M. Kamperman, F.J. DiSalvo, U. Wiesner, Nat. Mater. **7**(3), 222–228 (2008)
14. E.J.W. Crossland, *Block Copolymer Patterning of Functional Materials* (University of Cambridge, Cambridge, 2008)
15. H.-Y. Hsueh, Y.-C. Huang, R.-M. Ho, C.-H. Lai, T. Makida, H. Hasegawa, Adv. Mater. **23**(27), 3041–3046 (2011)
16. H.-Y. Hsueh, R.-M. Ho, Langmuir **28**(22), 8518–8529 (2012)
17. T. Wagner, T. Sauerwald, C.-D. Kohl, T. Waitz, C. Weidmann, M. Tiemann, Thin Solid Films **517**(22), 6170–6175 (2009)
18. S. Ndoni, M.E. Vigild, R.H. Berg, J. Am. Chem. Soc. **125**(44), 13366–13367 (2003)
19. W. Ma, C. Yang, X. Gong, K. Lee, A.J. Heeger, Adv. Funct. Mater. **15**(10), 1617–1622 (2005)
20. T. Hashimoto, Y. Nishikawa, K. Tsutsumi, Macromolecules **40**(4), 1066–1072 (2007)
21. A. Urbas, M. Maldovan, P. DeRege, E. Thomas, Adv. Mater. **14**(24), 1850–1853 (2002)
22. H. Mao, M.A. Hillmyer, Soft Matter **2**(1), 57–59 (2006)
23. T. Thurn-Albrecht, R. Steiner, J. De Rouchey, C.M. Stafford, E. Huang, M. Bal, M. Tuominen, C.J. Hawker, T.P. Russell, Adv. Mater **12**(11), 787–791 (2000)
24. I. Vukovic, T.P. Voortman, D.H. Merino, G. Portale, P. Hiekkataipale, J. Ruokolainen, G. ten Brinke, K. Loos, Macromolecules **45**(8), 3503–3512 (2012)
25. E.J.W. Crossland, S. Ludwigs, M.A. Hillmyer, U. Steiner, Soft Matter **3**(1), 94–98 (2007)
26. M.I. Kim, T. Wakada, S. Akasaka, S. Nishitsuji, K. Saijo, H. Hasegawa, K. Ito, M. Takenaka, Macromolecules **42**(14), 5266–5271 (2009)
27. E.J.W. Crossland, M. Kamperman, M. Nedelcu, C. Ducati, U. Wiesner, D.M. Smilgies, G.E.S. Toombes, M.A. Hillmyer, S. Ludwigs, U. Steiner, H.J. Snaith, Nano Lett. **9**(8), 2807–2812 (2009)
28. S. Khlebnikov, H.W. Hillhouse, Phys. Rev. B **80**(11), 115316 (2009)

Chapter 2
Gyroid and Gyroid-Like Surfaces

This chapter gives an overview of the fascinating gyroid and gyroid-like surfaces, which are found in nature as intermaterial dividing surfaces. The single- and double-gyroid surfaces in particular will be discussed. In order to visualize these gyroidal structures the concept of a level surface will be introduced and applied.

2.1 Intermaterial Dividing Surfaces

The immiscible components of block copolymers separate into distinct domains on the microscale in response to thermodynamic driving forces, that will be discussed in Chap. 4. The interfaces separating these domains have approximately constant mean curvature

$$H = \frac{1}{2}(k_1 + k_2) = \text{const.}, \tag{2.1}$$

where k_1 and k_2 are the principal curvatures. These surfaces are called intermaterial dividing surfaces (IMDS). Beside a number of unconnected periodic arranged IMDSs (e.g. spheres, lamellae and cylinders), multicontinuous triply periodic surfaces exist which are of particular interest, see Chap. 1. A bicontinuous (tricontinuous) triply periodic IMDS separates space into two (three) regions, forming connected networks which are periodic in three independent directions, such as the double primitive structure, discovered in a polymer/ceramic precursor composite [2, 3]. However, some of these structures were reevaluated as deformed double-gyroids [4]. The only bicontinuous IMDS in the phase diagram of pure diblock copolymers is the orthorhombic *Fddd* network and the only tricontinuous is the double-gyroid [5]. Copolymers containing more than two chemically distinct and incompatible block materials can adopt morphologies with more than three continuous phases, including the pentacontinuous core-shell double-gyroid discovered in ABC triblock copolymers [6, 7]. In other

---

Parts of this chapter are taken from my master's thesis [1].

M. R. J. Scherer, *Double-Gyroid-Structured Functional Materials*,
Springer Theses, DOI: 10.1007/978-3-319-00354-2_2,

self-organizing systems the bicontinuous single-gyroid microdomain structure can be found. Cubic phases with space group symmetry $I4_132$ (double-gyroid) and $Im\overline{3}m$ (double primitive) have also been identified in thermotropic liquid crystals [8].

All IMDSs can be approximated by constant mean curvature (CMC) surfaces that minimize interfacial area subject to a volume (or volume fraction) constraint. This fact is not surprising, since the microdomain structure is the result of the balance between chain-stretching and the interfacial energies, where the latter term seeks to minimize surface area and thus favors CMC surfaces.

Since analytical expressions for only a few continuous triply periodic CMC surfaces are known (e.g. the Enneper-Weierstrass parameterization of the single-gyroid minimal surface with $H = 0$ and a volume fraction of 50 % [9]), these surfaces are typically modeled with the help of level surfaces.

2.2 Level Surfaces

Level surfaces are represented by functions $F : \mathbb{R}^3 \rightarrow \mathbb{R}$ of points $(x, y, z) \in \mathbb{R}^3$, which satisfy the equation

$$F(x, y, z) = t, \tag{2.2}$$

where t is a constant [10]. The form of the surface is controlled by $F(x, y, z)$, while the parameter t determines the volume fraction of the regions that are separated by the surface.

Level surfaces of all IMDSs are defined by

$$F(x, y, z) = \sum_{hkl} |F(hkl)| \cdot \cos\left(\frac{2\pi}{L}(hx + ky + lz) - \alpha_{\text{hkl}}\right), \tag{2.3}$$

where $F(hkl)$ denotes the structure factor amplitude, reflecting the symmetry of the structure, α_{hkl} the phase angle and L the cubic unit cell edge length [11]. The set of allowed hkl values of the Fourier components of the structure and the Fourier series representation of the space group can be found in the *International tables of X-ray crystallography*. It was found that the generated level surface is virtually independent of the number of coefficients of the Fourier series, thus taking only the lower orders hkl values into account provides a sufficiently close approximation to the true CMC surface [11].

2.3 Single-Gyroid

The single-gyroid (SG) IMDS with $I4_132$ (No. 214) symmetry was first discovered in 1967 by Luzzati et al. as a cubic phase occurring in strontium soap surfactants and in pure lipid-water systems [12, 13]. In 1970, Schoen identified the minimal

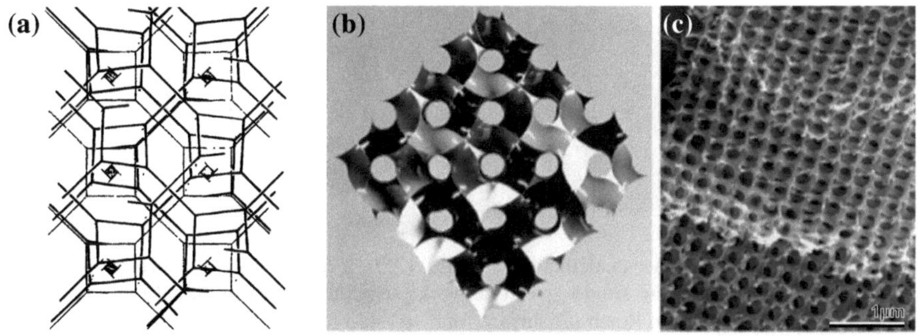

Fig. 2.1 The single-gyroid or Schoen G surface. **a** Drawing by Luzzati et al. from 1970 showing the two mutually interwoven networks separated by the G surface (Reprinted by permission from Macmillan Publishers Ltd: Nature, Luzzati et al. [12], Copyright 2013). **b** Plastic model of the gyroid minimal surface by Schoen from 1970 (Reprinted from [14]). **c** Scanning electron micrograph showing the porous cuticular single-gyroid of a *Callophrys rubi* wing scale. (Reprinted from Schröder-Turk et al. [18], Copyright 2013, with permission from Elsevier)

gyroid, therefore also called the Schoen G surface, as one of a further 17 examples of intersection-free triply periodic minimal surfaces (TPMS), in addition to the five cases already known, and gave a mathematical description [14].

The SG IMDS has been discovered in the cuticular structure of butterfly wing scales [15, 16]. The structure is made of chitin and air serving as biological photonic crystal to produce beautiful iridescent colors [16–18]. More recently, the mitochondria located at the inner segment of the retinal cones of tree shrew was found to contain a multi-layer gyroid membrane arrangement similar to the SG structures observed in butterflies. The highly order structure is believed to potentially function either as a multi-focal lens, as an angle-independent interference ultraviolet filter, or as a waveguide photonic crystal [19] (Fig. 2.1).

The single-gyroid of constant mean curvature can be approximated by a level surface. Although this approach does not result in CMC surfaces, it is extremely useful for simulations. Inserting the lowest allowed hkl values of the $I4_132$ space group ($|F(110)| = |F(\bar{1}10)| = 1$ and $\alpha(110) = \alpha(\bar{1}10) = \pi/2$) in Eq. (2.3) provides an unique equation for the single-gyroid level surface [11, 20]

$$F_{SG}(x, y, z) = t \tag{2.4}$$

$$\sin\left(\frac{2\pi}{L}x\right)\cos\left(\frac{2\pi}{L}y\right) + \sin\left(\frac{2\pi}{L}y\right)\cos\left(\frac{2\pi}{L}z\right) + \sin\left(\frac{2\pi}{L}z\right)\cos\left(\frac{2\pi}{L}x\right) = t, \tag{2.5}$$

where L is the cubic unit cell length.

The possible values for t and the resulting structures are discussed below:

- $t = 0$:
  The resulting level surface closely approximates the minimal gyroid or gyroid minimal surface discovered by Schoen, see Fig. 2.2a. Minimal surfaces have the

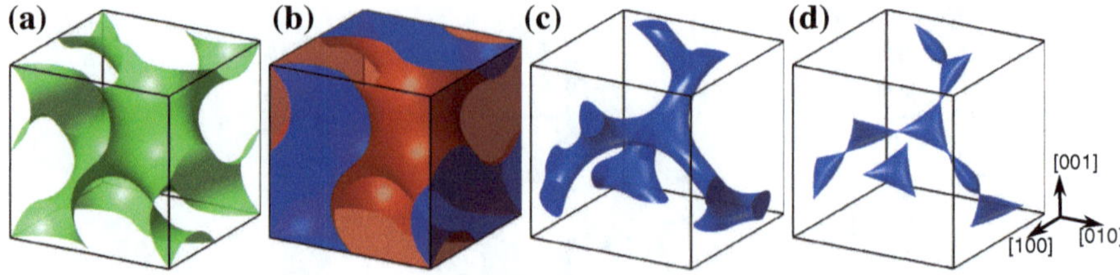

Fig. 2.2 Single-gyroid surfaces calculated using Eq. (2.5). **a** Minimal gyroid surface for $t = 0$. **b** Two helical interpenetrating single-gyroid networks separated by the minimal gyroid surface. **c** Single-gyroid for $t = 1.3$. **d** Pinch-off surface for $t = 1.413$

property that their mean curvature is zero everywhere ($H = 0$) and thus are a subset of CMC surfaces. For this special surface, as mentioned earlier, an analytical expression is known. Using the Enneper-Weierstrass representation, which provides an equation for the coordinates of minimal surfaces, the exact parametrization of the triply periodic gyroid minimal surface can be derived [9].

The minimal gyroid surface divides space into two helical regions or networks, each claiming a volume fraction of 50 %, see Fig. 2.2b. The surface embeds an inversion center which interchanges the two sides of the surface and also the two regions which it partitions. The separated networks are enantiomorphic, which means that one network is left-handed, the other one right-handed [9].

- $0 < |t| \leq 1.413$:
  An increasing $|t|$ is related to a monotonic increase/decrease in the volume of each of the two related networks and leads to an increase in the absolute mean curvature $|H|$, see Fig. 2.2c. Correlated to the change in volume fraction is the growing offset of the surface from the $t = 0$ base surface. For t, the surface on one side of the minimal surface ($t = 0$) with a certain offset, while for $-t$ the surface lies on the opposite side with the same offset. This behavior will become clearer when discussing the double-gyroid.
  For $t = 1.413$, the so-called pinch-off surface is generate, see Fig. 2.2d. Experimentally this IMDS is generated when the voided minority networks are replicated by either chemical vapor or atomic layer deposition techniques [21].
- $1.413 < |t| \leq 1.5$:
  For these t values, the SG surface is no longer connected and will be neglected in the following discussion.
- $|t| > 1.5$:
  The SG disappears, since no point $(x, y, z) \in \mathbb{R}^3$ fulfils Eq. (2.5).

2.4 Double-Gyroid

In 1986, Thomas et al. were the first to discover the double-gyroid IMDS in multiarm star diblock copolymers of poly(isoprene) and poly(styrene), but incorrectly identified the new morphology, lying between the lamellar and cylindrical phase, as the double diamond [22]. Eight years later, in 1994, two groups independently characterized the IMDS observed in a low molecular diblock copolymer melt and in a diblock copolymer blend correctly as DG with $Ia\overline{3}d$ (No. 230) symmetry [23, 24]. These results suggested a reexamination of the star diblock structure with improved X-ray measurements, which led to a reevaluation to DG [25]. Thereafter, the double-gyroid IMDS has been observed in various diblock copolymers [26, 27], in nanoporous silica films [28], in ABA triblock copolymers [29, 30], and in ABC triblock copolymers where the A and C block have identical volume fractions [31].

The simplest method to construct the DG surface is by generating two SG level surfaces with $0 < t \leq 1.413$,

$$F_{SG}(x, y, z) = \pm t, \tag{2.6}$$

which defines a so-called double level surface. By squaring Eq. (2.6) the (single) level surface equation (see Eq. (2.2)) of the DG can be obtained

$$F_{DG}(x, y, z) = (F_{SG}(x, y, z))^2 = t^2 = t'. \tag{2.7}$$

Another way to derive the DG level surface function would have been to use Eq. (2.3) with $Ia\overline{3}d$ as the generating space group, which delivers the same result [10, 11].

In the interval $0 < t \leq 1.413$ the DG level surface effectively divides space into three embedded continuous subvolumes. This phase which is located between the two SG surfaces, centered on the minimal gyroid surface, forms the so-called matrix phase through which the two networks run. Each one of the three networks is periodic in all three principle directions. The matrix phase volume increases with increasing values of t, and thereby gradually reducing the strut diameters of the two enclosed networks.

The matrix is also called the majority phase, because it takes up about 63–66 % of the volume in DG morphologies formed by diblock polymers, and accordingly the networks together are called the minority phase. An unit cell with cubic symmetry $Ia\overline{3}d$ showing the matrix phase together with the two minority networks is presented in Fig. 2.3a. To illustrate that the two distinct intertwined networks are non-intersecting, an arrangement of four unit cells with the matrix phase removed and the two networks colored blue and red is shown in Fig. 2.3b.

When using porous double-gyroid morphologies for nanotechnology applications by selectively removing either the matrix phase or the two networks, the surface area per bulk volume is an important parameter [32]. The double-gyroid's surface area A thereby depends on the network phase volume fraction ϕ. The two parameters necessary to determine the surface area per unit bulk volume were calculated numerically for the different values of t in the following fashion: The surface area was determined by triangulating the DG surface and then adding the area of the individual triangles up, while the volume of the networks was calculated by subdividing the unit cell in small cubes and then counting the number of cubes enclosed by the surfaces of the networks. Fig. 2.4a shows the almost linear dependence of the volume fraction ϕ of

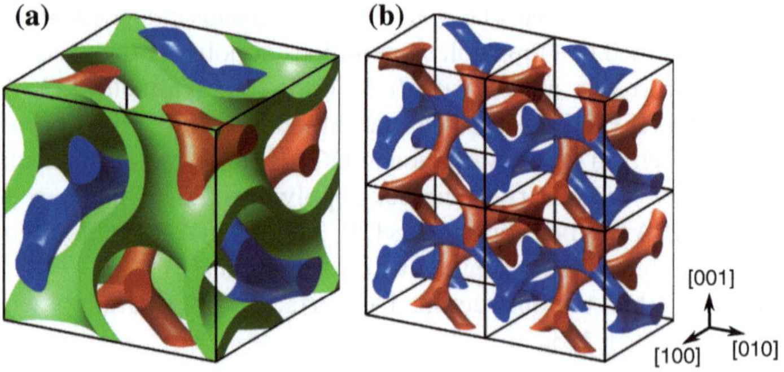

Fig. 2.3 **a** The double-gyroid cubic unit cell showing the matrix phase (*green*) and the two networks (*blue* and *red*). **b** Four cubic unit cells of the double-gyroid network with a volume fraction of 12 % ($t = \pm 1.3$)

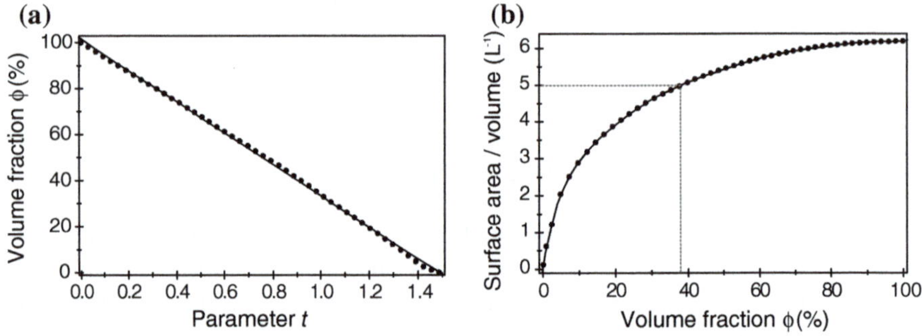

Fig. 2.4 **a** Volume fraction ϕ of the double-gyroid network phase as function of the parameter t fitted with $\phi(t) = 101.5 - 68.1 \cdot t$. **b** Specific surface area to bulk volume ratio of the double-gyroid plotted versus the network phase volume fraction in units of the cubic unit cell dimension L

DG networks on the parameter t. The surface area per unit bulk volume is plotted against the volume fraction of the networks in Fig. 2.4b.

When using structured materials in nanotechnology applications, it is important to know how robust these structures are. The weak points or cleavage planes of a three-dimensional porous structure often coincide with particular two-dimensional cross-sectional planes with a minimal area fraction of the nanostructured material. In order to determine the planes with minimal area fraction of the majority and minority voided double-gyroid networks a MATLAB simulation was used. The simulation numerically calculates the area fraction $A_{\mathrm{hkl}}(d)$ of the 65 % volume percent matrix network parallel to various crystallographic (hkl) planes as a function of the distance d from the point of origin to the intersection point of the plane with the [111] unit cell body diagonal. The results are shown in Fig. 2.5, where $A_{\mathrm{hkl}}(d)$ is plotted against d for the lowest values of $h^2 + k^2 + l^2$. Only for planes parallel to (211) and (110), the area fraction varies significantly around the mean value of 0.65 for different values of d. Whereas the area fraction of the other shown crystallographic planes

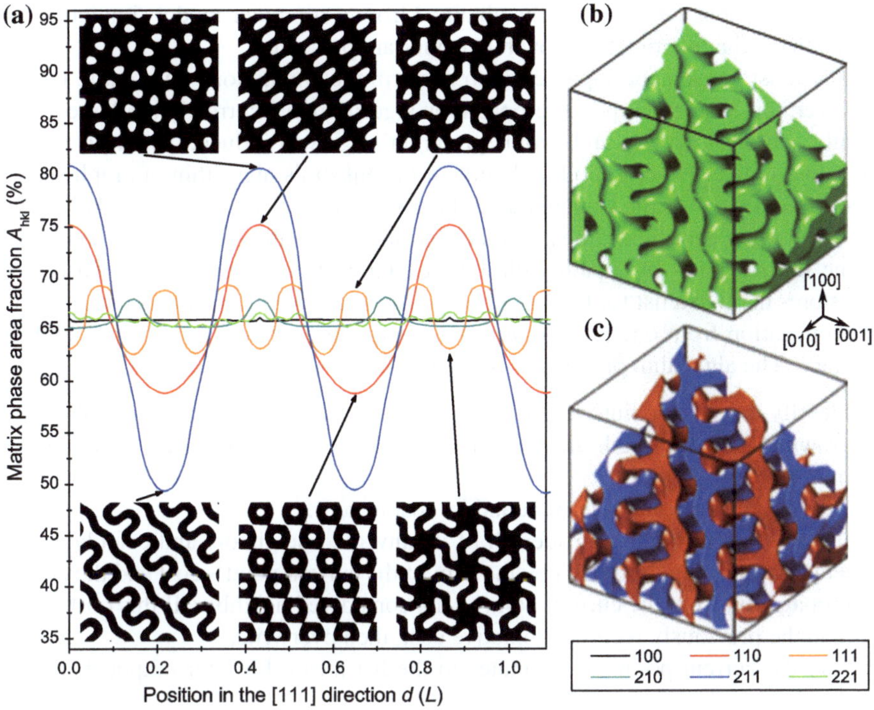

Fig. 2.5 A majority phase volume fraction of 0.65 was used for the presented simulations. **a** The area fraction $A_{hkl}(d)$ of the matrix phase's cross-sectional cut parallel to various crystallographic hkl planes is plotted against d, where d denotes the distance from the unit cell point of origin to the intersection point of the cross-section with the [111] unit cell body diagonal [33]. The two-dimensional patterns of (110), (111), and (211) plane possessing an extremal area fraction are shown. The matrix phase is colored in black. **b** Matrix and **c** network phase showing the (211) double wave pattern. The frame outlines the volume of 3×3 unit cells

and those with higher $h^2 + k^2 + l^2$ values hardly depends on d [33]. Thus, especially the cross-sections parallel to the (211) and also the (110) plane with minimal area fraction might serve as predetermined breaking points for double-gyroid majority network structured materials, while the planes with maximal area fraction might act as cleavage planes for minority network structures.

The complex triply periodic branching of gyroidal surfaces defines a fascinating three-dimensional maze. Finding the shortest path connecting two points lying within the maze is interesting from both a theoretical and an experimental point of view. A spherical growth front nucleated at a single site, the starting point, and propagated through this voided double-gyroid maze will find the shortest path to any given end point. Additionally, the propagating growth front will disclose any preferential growth directions through the maze. Experimentally, this can be tested by electroplating an amorphous or highly polycrystalline metal and limiting the number of nucleation sites, see Chap. 6. Since an analytical solution to this problem seems impossible, here, the results obtained from a computer model simulating the templated and nontemplated growth behavior are presented.

The most convenient approach is to perform the simulations on a cubic lattice, however, this requires a special propagation algorithm that correctly implements the locally spherical growth and avoids preferential growth induced by the nature of the cubic lattice. Here, a modified version of Dijkstra's algorithm, a graph search algorithm that solves the single-source shortest path problem, is employed [34]. For a given starting coordinate, the modified algorithm finds all points that are connected to the starting point by a path with lowest total cost equal to a given total cost C and stores them in a list for the subsequent visualization and analysis. Here, the cost of propagation from one point to a nearest neighbor is defined by their Euclidean distance. The algorithm propagates in the following way:

1. Identify the set of points for the cubic lattice of interest that are accessible to the growing front and mark them as unvisited by assigning a tentative cost value of infinity.
2. Set the starting point as current and its cost to zero.
3. For the current point, consider all of its unvisited neighbors and calculate their tentative cost which is given by the sum of the current point's cost and the cost of propagation from the current point to the considered neighbor. If this cost is less than the previously recorded tentative cost, then overwrite.
4. Add the current point's coordinates to the list meant for later output if its cost equals C.
5. Mark the current point as visited and remove it from the unvisited set.
6. Determine which unvisited point has the lowest cost and set it as the next current point. If the cost does not exceed C, then return to step 3.

Increasing the set of points that are considered as neighbors gives an improved approximation to a spherical growth. However, increasing the distance between the current point and neighbor can lead to omitting a maze obstacle and more time intensive calculations. Here, points lying within a radius of $a\sqrt{11}$ are considered as neighbors, where a is the cubic lattice's repetition length. To test if the model approximates a spherical growth the simulation was run in absence of a maze or scaffold, see Fig. 2.6.

Since the double-gyroid has a cubic unit cell it is sufficient to only consider points with positive coordinates and obtain the full three-dimensional propagation map by mirroring along the coordinate planes. Further, since the two interwoven double-gyroid minority networks share an inversion center that coincides with the unit cell center it is sufficient to restrict the simulation to one network. Both these measures help to reduce the calculation complexity, nevertheless, considering a volume of $7\times7\times7$ gyroid unit cells with a resolution of 60 along each coordinate axis leads to a cubic lattice with a total of 74 million cubes. The results of simulations for a double-gyroid with the voided minority phase having a volume fraction of 12 % are presented in Figs. 2.6 and 2.7.

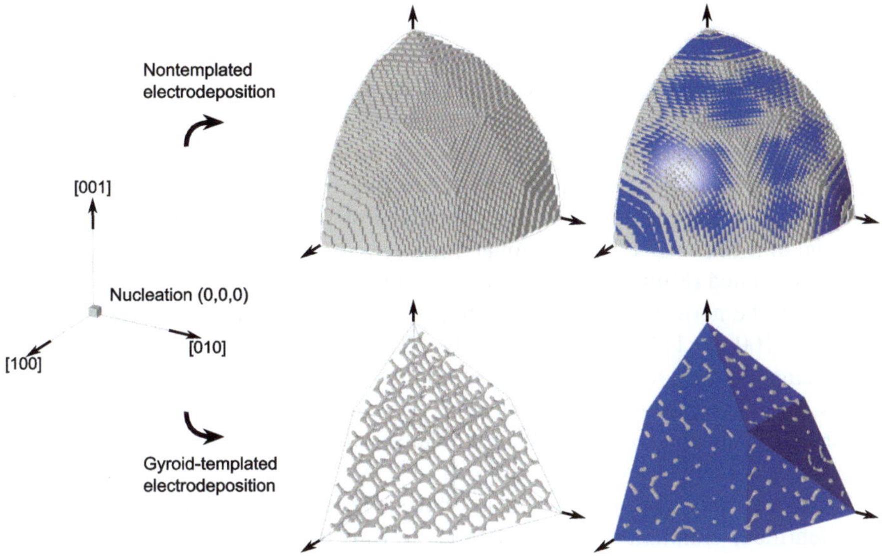

Fig. 2.6 Computer simulation of nontemplated and gyroid-templated amorphous/highly polycrystalline electroplating. In the case of nontemplated deposition a spherical growth front is obtained, as expected. Interestingly, the gyroid-templated simulation reveals preferential growth directions resulting in distinct facets

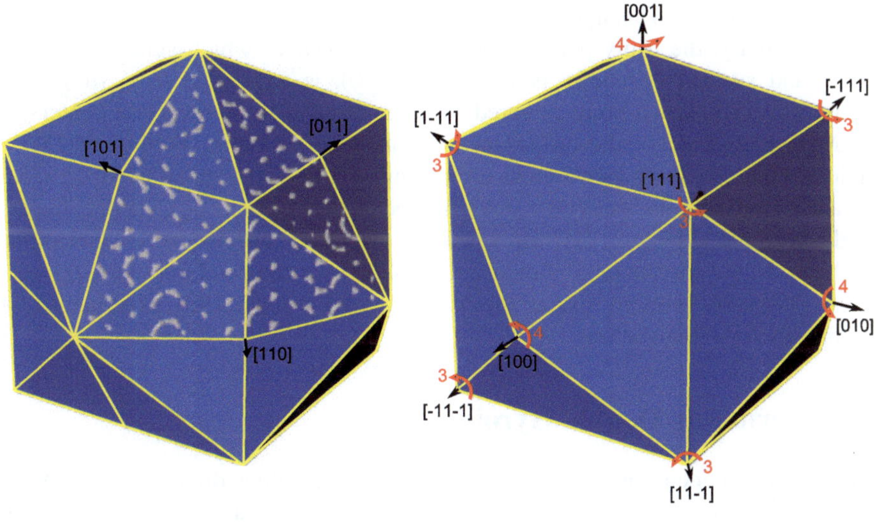

Fig. 2.7 Growth map for a double-gyroid with a volume fraction of 12 %. For this volume fraction the <110> vertices vanish since they lie on a straight line connecting two members of the <111> set. Thus, the <100> and <111> vertices are surrounded by 4 and 6 facets, respectively. The order of rotational symmetry is given in *red*

Preliminary analysis identified the <111> direction to have the fastest growth rate, however, it is still slower than the nontemplated growth. The growth speed shows a local maximum along the <100> directions, while the slowest propagation was found to exist along the <110> direction. The relative growth rate <110>:<100>: <111> is approximately $\sqrt{2}$:$\sqrt{2.5}$:$\sqrt{3}$, which means that the < 110 > vertices lie in very close vicinity of a straight line connecting two members of the <111> set and thus do not appear as such. A comparison of this simulation and one with a gyroidal unit cell rotated by 45° around the z-axis gave the same results, confirming that the obtained results are not an artifact of the propagation algorithm.
Increasing the networks' volume fraction to 35 % changes the relative growth rate <110>:<100>:<111> to approximately $\sqrt{2.4}$:$\sqrt{3}$:$\sqrt{3}$ with the speed in fast directions being identical to nontemplated speed. In this case the <110> directions form vertices, however, they still lie in close vicinity of the straight line connecting two members of the <111> set, making the detection of these vertices and the corresponding facet by means of SEM difficult.

These are only preliminary results, but since the extremal growth directions are identified, simulation restricted to these direction can be performed with a higher accuracy and resolution. Implementing these simulations should be relatively straight-forward and will help to identify the exact growth speed along these directions as a function of the networks' volume fraction. Nevertheless, this computer model suggests that the periodicity of templates with large gyroidal single domains will result in a crystal-like appearance of an electroplated amorphous or highly polycrystalline material. This hypothesis will be tested in Chap. 6.

In summary, the triply periodic double-gyroid IMDS, which generating space group and surface symmetry is $Ia\overline{3}d$ (No. 230), is embedded (free from self-intersection) and has a body centered cubic unit cell with an inversion center. As an IMDS every point on the surface has constant mean curvature. Further, the IMDS divides space into three subvolumes each being continuous, forming two interwinding and enantiomorphic networks, one left-handed, the other one righthanded. Importantly, the three subvolumes are triply periodic and interconnected, which means that on their own they define three-dimensionally self-supporting and freestanding structures. This latter property makes the double-gyroid so interesting as structure for nanotechnology applications.

2.5 Alternating Double-Gyroid

The two interpenetrating networks defining an alternating double-gyroid IMDS posses different volume fractions and as such the IMDS has $I4_132$ symmetry, as discovered in ABC triblock copolymers [31]. However, in the case that the A and C block have identical volume fractions the IMDS has $Ia\overline{3}d$ symmetry, but is still considered to be an alternating double-gyroid since the two networks are made of chemically different materials and the structure as such only has $I4_132$ symmetry. The points (x, y, z) fulfilling Eq. (2.8) define the alternating double-gyroid surface

$$F_{SG}(x, y, z) = \begin{cases} t_1 \\ -t_2 \end{cases}, \tag{2.8}$$

for $t_1 \neq t_2$ and $0 < t_{1,2} \leq 1.413$.

2.6 Core-Shell Double-Gyroid

The core-shell IMDS was first discovered in ABC triblock copolymers where two chemically identical interpenetrating double-gyroid networks are comprised of cores of C encased in shells of B and embedded in a matrix of A [6, 7, 35, 36]. Thus, the core-shell double-gyroid IMDS in these triblock copolymers is pentacontinuous. One could envision, that a copolymer with four or more chemically distinct blocks could also form a version of the core-shell gyroid with more than one shell.

The values of (x, y, z) that solve the following 'quadruple' level surface equation generate the core-shell gyroid surface

$$F_{SG}(x, y, z) = \begin{cases} \pm t_1 \\ \pm t_2 \end{cases}, \tag{2.9}$$

for $t_1 \neq t_2$ and $0 < t_{1,2} \leq 1.413$. The core-shell double-gyroid is a superposition of two double-gyroids with different volume fractions, sharing the same inversion center which results in $Ia\overline{3}d$ symmetry. The symmetry is preserved also when taking the core's and the shell's materials into consideration, since both of their two networks consist out of chemically identical material.

2.7 Gyroid-Like Surfaces

Two more gyroid-like IMDSs were discovered in the phase diagram of copolymers, namely the orthorhombic *Pnna* (No. 52) and the orthorhombic *Fddd* (No. 70) network, see Fig. 2.8. Bailey and colleagues were the first to observe the *Fddd* IMDS in copolymer systems, namely PI-*b*-PS-*b*-PEO. The drawings of the *Fddd* published in [40] suggest a 'core-shell *Fddd*' morphology, while in a later publication [31] an 'alternating *Fddd*' is presented. After the discovery of this noncubic triply periodic morphology in triblock copolymers the orthorhombic *Fddd* was also found to be adopted by diblock copolymers [5, 37]. The *Fddd* IMDS formed by diblock copolymers separates space into two regions that form connected networks which are periodic in three independent directions. Hence, it is considered to be bicontinuous and triply periodic just like the single-gyroid surface.

The first to prove the existence of the orthorhombic *Pnna* IMDS in copolymer melts were Cochran et al. [41]. A triblock copolymer melt was sheared in order

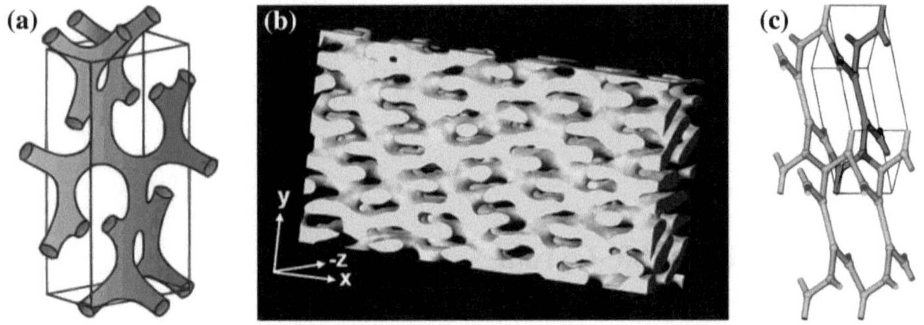

Fig. 2.8 **a** Unit cell of the *Fddd* structure found in diblock copolymers. (Reprinted with permission from Kim et al. [37]. Copyright 2013 American Chemical Society) **b** Reconstructed three-dimensional image of a poly(styrene)-*b*-poly(isoprene) *Fddd* structure obtained by transmission electron microtomography. (Reproduced from Jung et al. [38] with permission of The Royal Society of Chemistry) **c** Ball and stick model of the *Pnna* (No. 52). (Reproduced from Bluemle et al. [39] with permission of The Royal Society of Chemistry)

to induce a network-to-network transition from *Fddd* to *Pnna*. Subsequent to the discovery of the shear-induced *Pnna* morphology, this structure was also found to exist in tetrablock copolymers [39].

References

1. M.R.J. Scherer, Nanostructured Materials via Self-Assembled Templates, University of Konstanz (2009)
2. A.C. Finnefrock, R. Ulrich, A.D. Chesne, C.C. Honeker, K. Schumacher, K.K. Unger, S.M. Gruner, U. Wiesner, Angew. Chem. Int. Ed. **40**(7), 1207–1211 (2001)
3. B.C. Garcia, M. Kamperman, R. Ulrich, A. Jain, S.M. Gruner, U. Wiesner, Chem. Mater. **21**(22), 5397–5405 (2009)
4. G.E.S. Toombes, A.C. Finnefrock, M.W. Tate, R. Ulrich, U. Wiesner, S.M. Gruner, Macromolecules **40**(25), 8974–8982 (2007)
5. M. Takenaka, T. Wakada, S. Akasaka, S. Nishitsuji, K. Saijo, H. Shimizu, M.I. Kim, H. Hasegawa, Macromolecules **40**(13), 4399–4402 (2007)
6. T. Goldacker, V. Abetz, Macromolecules **32**(15), 5165–5167 (1999)
7. T.A. Shefelbine, M.E. Vigild, M.W. Matsen, D.A. Hajduk, M.A. Hillmyer, E.L. Cussler, F.S. Bates, J. Am. Chem. Soc. **121**(37), 8457–8465 (1999)
8. X. Zeng, G. Ungar, M. Imperor-Clerc, Nat. Mater. **4**(7), 562–567 (2005)
9. P.J.F. Gandy, J. Klinowski, Chem. Phys. Lett. **321**(5–6), 363–371 (2000)
10. M. Wohlgemuth, N. Yufa, J. Hoffman, E.L. Thomas, Macromolecules **34**(17), 6083–6089 (2001)
11. H.G. von Schnering, R. Nesper, Z Phys. B Con. Mat. **83**(3), 407–412 (1991)
12. V. Luzzati, P.A. Spegt, Nature **215**(5102), 701–704 (1967)
13. V. Luzzati, A. Tardieu, T. Gulik-Krzywicki, E. Rivas, F. Reiss-Husson, Nature **220**(5166), 485–488 (1968)
14. A.H. Schoen, Infinite periodic minimal surfaces without self-intersections, NASA Technical Note D–5541 (1970).
15. K. Michielsen, D. Stavenga, J.R. Soc, Interface **5**(18), 85–94 (2008)

16. P. Vukusic, J.R. Sambles, Nature **424**(6950), 852–855 (2003)
17. M. Maldovan, A.M. Urbas, N. Yufa, W.C. Carter, E.L. Thomas, Phys. Rev. B **65**(16), 165123 (2002)
18. G. Schröder-Turk, S. Wickham, H. Averdunk, F. Brink, J. Fitz Gerald, L. Poladian, M. Large, S. Hyde, J. Struct. Biol. **174**(2), 290–295 (2011)
19. Z. Almsherqi, F. Margadant, Y. Deng, Interface Focus (2012)
20. Y. Grin, R. Nesper, Z. Krist. **226**(8), 692–710 (2011)
21. J.H. Moon, Y. Xu, Y. Dan, S.-M. Yang, A.T. Johnson, S. Yang, Adv. Mater. **19**(11), 1510–1514 (2007)
22. E.L. Thomas, D.B. Alward, D.J. Kinning, D.C. Martin, D.L. Handlin, L.J. Fetters, Macromolecules **19**(8), 2197–2202 (1986)
23. D.A. Hajduk, P.E. Harper, S.M. Gruner, C.C. Honeker, G. Kim, E.L. Thomas, L.J. Fetters, Macromolecules **27**(15), 4063–4075 (1994)
24. M.F. Schulz, F.S. Bates, K. Almdal, K. Mortensen, Phys. Rev. Lett. **73**(1), 86 (1994)
25. D.A. Hajduk, P.E. Harper, S.M. Gruner, C.C. Honeker, E.L. Thomas, L.J. Fetters, Macromolecules **28**(7), 2570–2573 (1995)
26. G. Floudas, R. Ulrich, U. Wiesner, J. Chem. Phys. **110**(1), 652–663 (1999)
27. S. Ndoni, M.E. Vigild, R.H. Berg, J. Am. Chem. Soc. **125**(44), 13366–13367 (2003)
28. V.N. Urade, T.-C. Wei, M.P. Tate, J.D. Kowalski, H.W. Hillhouse, Chem. Mater. **19**(4), 768–777 (2007)
29. B.J. Dair, C.C. Honeker, D.B. Alward, A. Avgeropoulos, N. Hadjichristidis, L.J. Fetters, M. Capel, E.L. Thomas, Macromolecules **32**(24), 8145–8152 (1999)
30. A. Nykänen, M. Nuopponen, A. Laukkanen, S.-P. Hirvonen, M. Rytelä, O. Turunen, H. Tenhu, R. Mezzenga, O. Ikkala, J. Ruokolainen, Macromolecules **40**(16), 5827–5834 (2007)
31. T.H. Epps, E.W. Cochran, T.S. Bailey, R.S. Waletzko, C.M. Hardy, F.S. Bates, Macromolecules **37**(22), 8325–8341 (2004)
32. C.A. Lambert, L.H. Radzilowski, E.L. Thomas, Philos. Trans. R. Soc. London, Ser. A, **354**(1715), 2009–2023 (1996)
33. T. Hashimoto, Y. Nishikawa, K. Tsutsumi, Macromolecules **40**(4), 1066–1072 (2007)
34. E.W. Dijkstra, Numer. Math. **1**(1), 267–271 (1959)
35. A.J. Meuler, C.J. Ellison, M.A. Hillmyer, F.S. Bates, Macromolecules **41**(17), 6272–6275 (2008)
36. M. Stefik, S. Mahajan, H. Sai, T.H. Epps, F.S. Bates, S.M. Gruner, F.J. DiSalvo, U. Wiesner, Chem. Mater. **21**(22), 5466–5473 (2009)
37. M.I. Kim, T. Wakada, S. Akasaka, S. Nishitsuji, K. Saijo, H. Hasegawa, K. Ito, M. Takenaka, Macromolecules **41**(20), 7667–7670 (2008)
38. J. Jung, H.-W. Park, J. Lee, H. Huang, T. Chang, Y. Rho, M. Ree, H. Sugimori, H. Jinnai, Soft Matter **7**(21), 10424 (2011)
39. M.J. Bluemle, G. Fleury, T.P. Lodge, F.S. Bates, Soft Matter **5**(8), 1587–1590 (2009)
40. T.S. Bailey, C.M. Hardy, T.H. Epps, F.S. Bates, Macromolecules **35**(18), 7007–7017 (2002)
41. E.W. Cochran, F.S. Bates, Phys. Rev. Lett. **93**(8), 087802 (2004)

Chapter 3
Diblock Copolymer Synthesis

The aim of this first part of my study was to establish a reliable, robust, and scalable synthesis route to double-gyroid-forming diblock copolymers.

The gyroid-forming poly(4-fluorostyrene)-*b*-poly(D,L-lactic acid) (PFS-*b*-PLA) diblock copolymer, denoted MHI41, used by Crossland et al. was synthesized by Prof. Marc Hillmyer and Dr. Sabine Ludwigs in three steps: Firstly, hydroxy-terminated PLA was synthesized by a trimethylaluminium-catalyzed coordination-insertion ring-opening polymerization (ROP) of lactide initiated by benzyl alcohol [2]; secondly, the hydroxyl group was converted into a reversible addition fragmentation transfer (RAFT) agent by reaction with the acid chloride of S-1-dodecyl-S'-(α,α'-dimethyl-α''-acetic acid) trithiocarbonate; and finally, the PFS block was added via the controlled RAFT polymerization of 4-fluorostyrene [3].

In contrast, here a bifunctional initiator is employed and the polymerization order of the two blocks is inverted: In a first step, the styrene block is synthesized by atom transfer radical polymerization (ATRP); followed by the addition of lactide via the recently developed organocatalytic ring-opening polymerization, as depicted in Fig. 3.1 [4, 5]. This synthesis route reduces the involved steps and enables a simplified and time-efficient preparation of copolymers with different block compositions. Importantly, both polymerization techniques offer precise and robust control over the copolymer composition, which is an essential requirement to reliably target the double-gyroid's narrow location in phase space [6].

In addition, diblock copolymers with different 4-X-styrene blocks, denoted as PXS, containing 4-chlorostyrene, 4-bromostyrene, or mixtures of 4-fluorostyrene and styrene, were synthesized. In contrast to poly(4-fluorostyrene), polymers of 4-bromostyrene and 4-chlorostyrene are cross-linkable by ultraviolet irradiation [7–9]. Cross-linked polymers typically show better chemical resistance and thermal stability, which are indispensable properties when used as template material for challenging templating methods. Furthermore, replacing the expensive 4-fluorostyrene

Parts of this chapter were previously published [1].

M. R. J. Scherer, *Double-Gyroid-Structured Functional Materials*,
Springer Theses, DOI: 10.1007/978-3-319-00354-2_3,

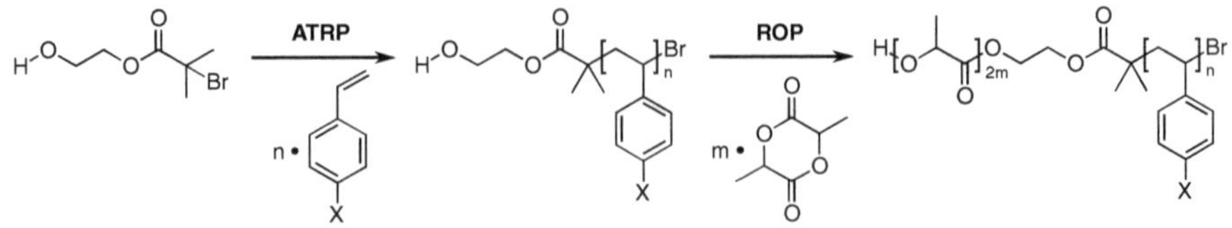

Fig. 3.1 Two-step synthesis of poly(4-X-styrene)-*b*-poly(lactic acid) employing ATRP, ROP, and the bifunctional initiator HEBIB (X=H, F, Cl, Br)

with styrene significantly reduces the total synthesis cost of double-gyroid-forming copolymers and makes the template preparation more environmentally sustainable.

3.1 Introduction to the Employed Polymerization Techniques

3.1.1 Atom Transfer Radical Polymerization

Since the development of the atom transfer radical polymerization (ATRP) by Wang and Matyjaszewski in 1995 it has become the most extensively studied controlled/living radical polymerization method [10, 11]. For a long time living (ionic) polymerization was the only method capable of preparing well-defined polymers, in terms of composition and molecular architecture, which is required by many high-value applications [12]. Enabling a comparable level of control over the synthesis of well-defined polymers with polydispersity indices as low as $M_w/M_n = 1.05$, combined with its simplicity and broad applicability made ATRP extremely popular [13]. ATRP also enables the fast synthesis of high molar mass polymers of acrylates, methacrylates, and vinyl chloride at 25 °C, making it highly attractive for industrial purposes [14]. Here, only a brief introduction on the vast research field of ATRP is presented, for more detailed information the reader is referred to one of the reviews on this topic [4, 15–17].

3.1.1.1 Fundamentals of ATRP

The radical or active species $P_n^\bullet$ with a degree of polymerization n is generated by a reversible redox reaction of an alkyl halide initiator or a dormant species P_nX with an activator, typically a low-oxidation-state metal/ligand complex Mt^zY/L, where X denotes the haloalkane species, z the metal's oxidation state, Y a possible counterion, and L the ligand (Fig. 3.2) [16]. This reversible redox process involves an one-electron oxidation of the Mt^zY/L complex with a concomitant abstraction of a (pseudo)halogen atom X from the dormant species P_nX. Thereby, the propagating radical $P_n^\bullet$ and the higher-oxidation-state metal/ligand complex $XMt^{z+1}Y/L$, which acts as deactivator, are formed [18–22].

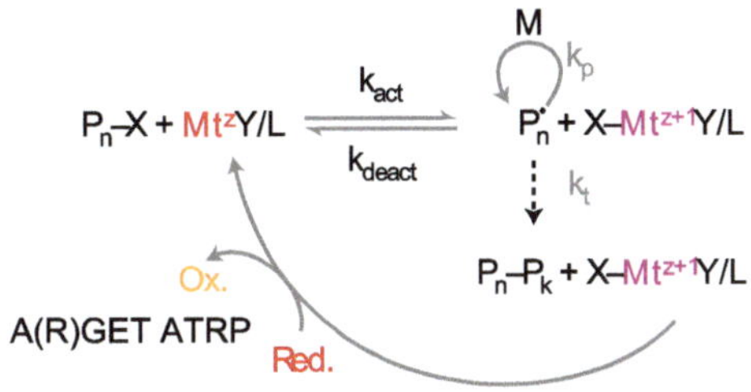

Fig. 3.2 Reaction scheme for ATRP. *Font color* indicates oxidation state changes (Adapted by permission from Macmillan Publishers Ltd: Nature Chemistry, Matyjaszewski et al. [16], Copyright 2013)

The described redox process occurs with the rate constant of activation k_{act} and deactivation k_{deact}. The activated polymer chains $P_n^•$ grow by the addition of the intermediate radicals to monomers M, with a propagation rate constant k_p. In a well-controlled ATRP, no more than a few percent of the polymer chains undergo termination, however, termination reactions do occur with a rate constant k_t mainly due to radical coupling (P_nP_k) and disproportionation [23]. When the activators are (re)generated by electron transfer from the high-oxidation-state, $XMt^{z+1}Y/L$, using a reducing agent, the polymerization is called A(R)GET ATRP, thereby reducing the amount of catalyst needed [18 22].

3.1.1.2 Kinetics, Molecular Weight and Polydispersity

Fast initiation and rapid reversible deactivation are the key to a successful ATRP, combing of a small number of terminated chains and an uniform chain growth mechanism. Thus, the polymerization rate R_p plays an important role and can be defined by the kinetic parameters and reagent concentrations

$$R_p = -\frac{d[M]}{dt} = k_p\,[M]\left[P_n^•\right] = k_p \cdot \frac{k_{act}}{k_{deact}} \cdot \frac{[M]\,[P_nX]\left[Mt^zY/L\right]}{\left[XMt^{z+1}Y/L\right]}. \tag{3.1}$$

R_p increases with temperature due to the increase of both the radical propagation rate constant k_p and the atom transfer equilibrium constant $K_{eq} = k_{act}/k_{deact}$ [16]. Usually higher k_p/k_t ratios and better control are observed at elevated temperatures due to a higher activation energy for the radical propagation than for the radical termination, but chain transfer and other side reactions may also become more pronounced. The optimal temperature depends mostly on the monomer, the catalyst, and the targeted molecular weight.

The polymer's degree of polymerization N prepared by a well-controlled ATRP can be predetermined for a monomer conversion factor c by the ratio of the initial amount of monomer and initiator

$$N = \frac{[M]}{[P_nX]_0} \cdot c. \tag{3.2}$$

The theoretical molecular weight distribution or polydispersity, PDI_{theo} for a well-controlled polymerization in the absence of significant chain termination and transfer, relates to the concentrations of initiator and deactivator, the rate constants of propagation and deactivation, and the monomer conversion in the following way [15]

$$PDI_{\text{theo}} = \frac{M_{\text{w}}}{M_{\text{n}}} = 1 + \frac{1}{N} + \frac{k_{\text{p}}}{k_{\text{deact}}} \cdot \frac{[\text{P}_n\text{X}]_0}{\left[\text{XMt}^{z+1}\text{Y/L}\right]} \cdot \left(\frac{2}{c} - 1\right). \tag{3.3}$$

Thus, for the same monomer a polymerization with faster deactivation will result in polymers with lower polydispersity, although at the cost of slower polymerization rates. This can be achieved with catalysts with a higher value of k_{deact} and/or with higher deactivator concentrations. Further, the polydispersity decreases with an increasing monomer conversion c and hence, is lower for high molecular weight polymers. However, the ATRP is normally terminated before complete monomer consumption, because in experiments the polydispersity starts to increase at nearly complete monomer conversion.

3.1.1.3 Components of ATRP

ATRP is a multicomponent system, composed of the monomer, an initiator with a transferable (pseudo)halogen, and a transition metal/ligand complex as catalyst, optionally a solvent or an additive, such as a reducing agent.

Monomers Although polymerized under identical conditions each monomer possesses its own unique atom transfer equilibrium constant K_{eq} and intrinsic radical propagation rate k_{p}. In the absence of any side reactions other than radical termination by coupling or disproportionation, the magnitude of the product $k_{\text{p}} \cdot K_{\text{eq}}$ determines the polymerization rate R_{p}. If R_{p} is too small ATRP will not occur or only very slowly. In contrast, if the equilibrium constant is too large this will lead to a significant occurrence of termination reactions because of a high radical concentration. However, R_{p} can be adjusted by choosing an adequate catalyst, catalyst concentration, temperature, optionally a solvent and additives to gain optimal polymerization control.

A variety of monomers have been successfully polymerized using ATRP [24]. Typical monomers include (meth)acrylates, (meth)acrylamides, acrylonitrile, and styrenes as shown in Fig. 3.3.

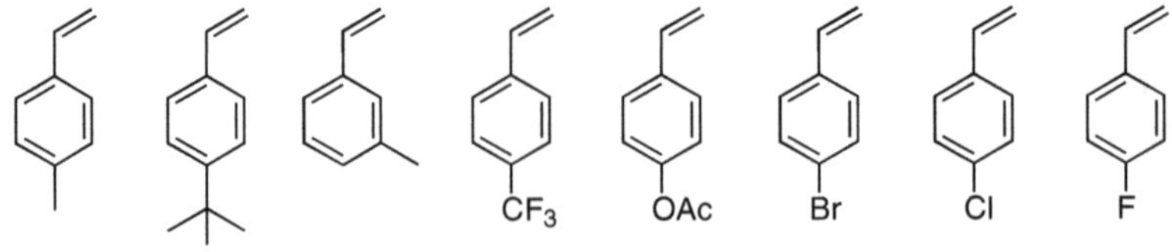

Fig. 3.3 Styrenic monomers polymerized by ATRP (Adapted with permission from Matyjaszewski et al. [4]. Copyright 2013 American Chemical Society)

Initiators As discussed previously, the amount of initiator in ATRP determines the number of growing polymer chains. To obtain well-defined polymers with narrow molecular weight distributions, the halide group, X, must rapidly and selectively migrate between the growing chain and the transition-metal complex. Bromine and chlorine were found to enable a precise control over the molecular weight. Further, iodine is used in contrast to fluorine because the C–F bond is too strong to undergo homolytic cleavage. The order of bond strength in the alkyl halides is R–Cl > R–Br > R–I [4]. In general, any alkyl halide with activating substituents on the R-carbon, such as aryl, carbonyl, or allyl groups can potentially be used as ATRP initiator. A variety of halogenated alkanes, benzylic halides, α-haloesters, α-haloketones α-halonitriles, and sulfonyl salides have been applied successfully [18–22]. Besides polyhalogenated compounds some pseudohalogens, specifically thiocyanates and thiocarbamates, have also been used in the polymerization of acrylates and styrenes. Figure. 3.4 illustrates that the activity of the initiator strongly depends on the transferable atom and the degree of substitution. Furthermore, the radical stabilizing group influences initiator activity. In general the following relation holds: CN > C(O)R > C(O)OR > Ph > Cl > Me and multiple functional groups may lead to a further increase in activity [4].

Catalysts A number of transition metal/ligand complexes have been applied in ATRP [4]. In order to generate growing radicals, the metal catalyst should undergo an electron transfer reaction with the abstraction of a (pseudo)halogen. Additionally, the oxidized catalyst should rapidly deactivate the propagating polymer chains to form the dormant species. Currently, the most efficient catalysts are complexes of late and middle transition metals (e.g. Cu, Fe, Ru, Rh, Ni, Pd, Co, Os, Re, Mo and Ti). Copper bromide and chloride are superior in ATRP in terms of versatility and cost [25]. However, catalytic activity and selectivity is strongly ligand dependent. Since most of the transition-metal salts are poorly soluble in organic media a main role of the ligand is to improve their solubility. Further, the ligand adjusts the redox potential of the metal center for appropriate reactivity and dynamics for the atom transfer. An appropriate ligand must be chosen in order to guarantee the following reaction conditions: Fast and quantitative initiation; a strongly shifted equilibrium between the alkyl halide and the transition metal toward the dormant species; fast deactivation

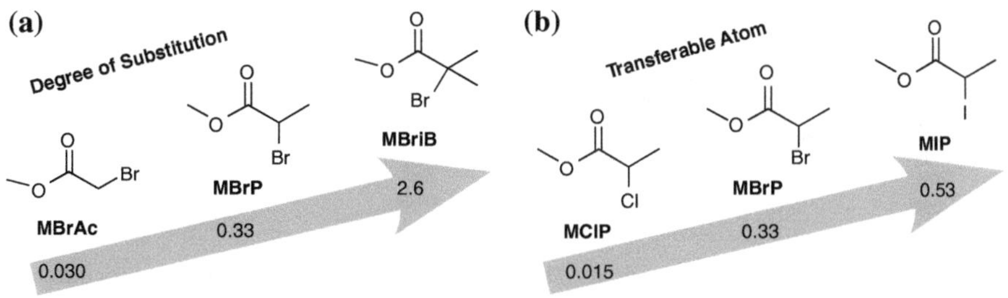

Fig. 3.4 Activity of various ATRP initiators. Values of k_{act} are given in units of $M^{-1}s^{-1}$ as determined in ATRP with CuX/PMDETA in MeCN at 35 °C (Adapted from Braunecker et al. [15], Copyright 2013, with permission from Elsevier)

of the active radicals; relatively fast activation of the dormant polymer chains; and finally, absence of side reactions or reduction/oxidation of the radicals [18–22].

Typically, phosphorus-based ligands are used to complex no-copper transition metals. Whereas, nitrogen-containing ligands have been used in copper- and iron-mediated ATRP, including 1,1,4,7,10,10-Hexamethyltriethylenetetramine (HMTETA), 2,2′-bipyridine (Bipy), N,N,N',N",N"-pentamethyldiethylenetriamine (PMDETA), and tris[2-(dimethylamino)ethyl]amine (Me_6TREN) [26, 27]. The activation rate constants for these ligands in copper bromide mediated ATRP and the proposed reaction mechanism of copper/2,2′-bipyridine are shown in Fig. 3.5.

Additives Copper-mediated ATRP is not affected by water, aliphatic alcohols, and polar compounds. This is a main advantage compared to the highly sensitive living ionic polymerization and enables the synthesis of well-defined functional polymers without the need for protection and deprotection of the functionalities [24]. However, ATRP is moderately sensitive to oxygen. Small amounts of oxygen can be scavenged by the catalyst, although the oxidation of the catalyst reduces the catalyst concentration and slows down polymerization. In contrast, some additives can accelerate ATRP, for example zero-valent copper(0) or tin(II)-2-ethylhexanoate, and thus enable the synthesis of high molecular weight polymers and copolymers with controlled architectures [21, 28]. Presumably, these additives reduce excess copper(II), generated mostly during the early stage of the polymerization through irreversible radical termination, to form in situ copper(I) by a simple electron transfer process.

It is also possible to start the reaction with copper(II) instead of copper(I) and then reduce this species with a reducing agent, thus termed activator generated by electron transfer (AGET ATRP) [18–20, 22]. When using reducing agents like tin(II)-2-ethylhexanoate, it is possible to catalyze the polymerization with parts per million amounts of copper [27]. Moreover, if a sufficient amount of zero-valent metal is present, the polymerization can be carried out without the removal of any oxygen or inhibitor, since the oxidized metal is regenerate by the zero-valent copper [29].

ATRP can be carried out either in bulk, in solution, or in a heterogeneous system. A solvent might be necessary when it is the aim to reach high levels of monomer

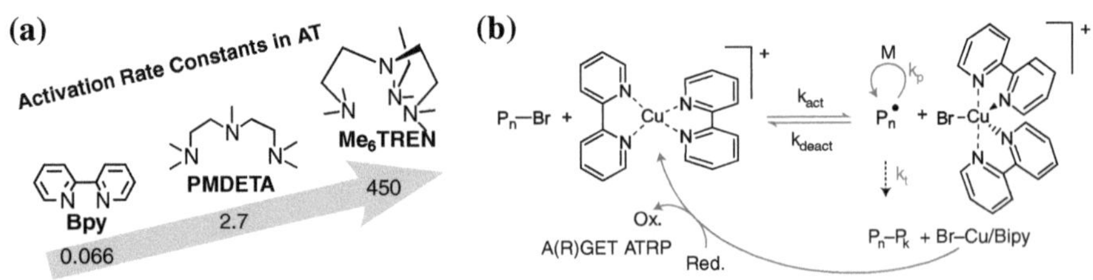

Fig. 3.5 Nitrogen-based ATRP ligands. **a** ATRP rate constants of activation in $M^{-1}s^{-1}$ for different ligands with ethyl-2-bromoisobutyrate as initiator in the presence of Cu^IBr in MeCN at 35 °C (Adapted from Braunecker et al. [15], Copyright 2013, with permission from Elsevier). **b** Proposed mechanism for copper/2,2′-bipyridine mediated ATRP (Adapted from Pintauer et al. [27] with permission of The Royal Society of Chemistry)

conversion, because the reaction solution may solidify when low monomer/polymer ratios are reached. Various solvents, such as benzene, toluene, anisole, diphenyl ether, and many others, have been used for different monomers. When using a solvent the possible interactions with the catalytic system must be considered.

3.1.1.4 Polymerization of Styrene and Substituted Styrenes

A majority of the literature on ATRP focuses on the synthesis of styrene and its derivatives with copper-based catalysts. One of the most extensively studied systems is the polymerization of styrene conducted with CuBr, nitrogen-based ligands, and alkyl bromides as initiator. Better molecular weight control is obtained at low temperatures, presumably due to a lower contribution of thermal self-initiation during the early stage of the polymerization. For example, the reaction temperature can be lowered to 80–90 °C when efficient catalysts, such as CuBr/PMDETA, are used [19–22, 30].

Numerous studies on kinetics, molecular weight, and polydispersity of styrene under different polymerization conditions were reported [31, 32]. In contrast, for substituted styrenes only a few reports exist. Qui et al. reported the controlled polymerization of various substituted styrenes by ATRP, among them 4-fluorostyrene [30]. ATRP of the styrene derivatives was performed in diphenyl ether at 110 °C. It was found, that monomers with electron withdrawing substituents polymerize faster than those bearing electron donating substituents in the order of 3-CF_3, 4-CF_3 > 4-Br, 4-Cl > 4-F, 4-H > 3-Me > 4-OMe > 4-Me > 4-CMe_3, as illustrated in Fig. 3.6a [30]. Another feature of substituted styrene ATRP is that polymers bearing electron withdrawing groups have lower polydispersities than those bearing electron donating groups, see Fig. 3.6b.

Becker et al. synthesized both blocks of poly(4-fluorostyrene)-*b*-poly(methyl acrylate) and poly(methyl acrylate)-*b*-poly(4-fluorostyrene) diblock copolymers

(a)
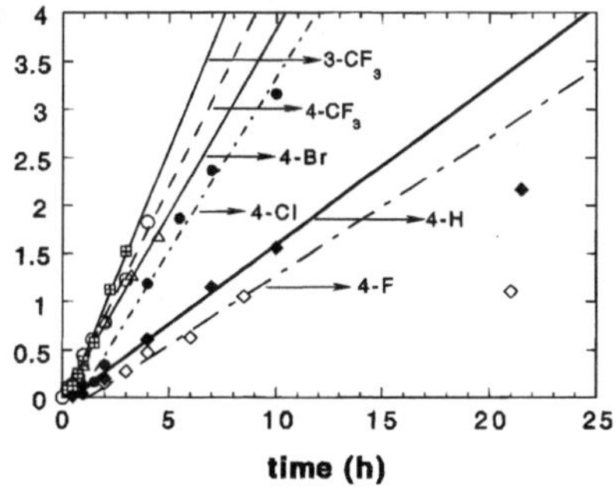

(b)
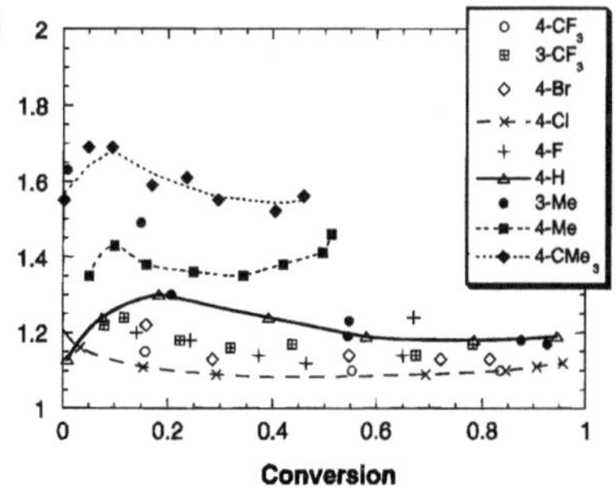

Fig. 3.6 ATRP of substituted styrenes. **a** Kinetic plot and **b** polydispersities of polymerizations performed in diphenyl ether at 110 °C with an initial monomer concentration of $[M]_0 = 4.37\,M$ and $[M]_0 : [1-PEBr]_0 : [CuBr]_0 : [bipy]_0 = 100 : 1 : 1 : 3$ (Reprinted with permission from Qui et al. [30]. Copyright 2013 American Chemical Society)

under ATRP conditions at 91–97 °C in toluene with CuBr/PMDETA as catalyst [33]. The poly(4-fluorostyrene) blocks typically had molecular weights up to 12.5 kg mol^{-1} and $M_w/M_n = 1.08$.

Bucholz et al. demonstrated the synthesis of poly(D,L-lactic acid)-*b*-poly(pentafluorostyrene) diblock copolymers using Cu(I)Br/bipyridine in a xylene/anisole mixture and a PLA-Br macroinitiator at 125 °C [34]. The pentafluorostyrene blocks typically had molecular weights up to 37 kg mol^{-1} and a polydispersity index of 1.11. A very elegant way to polymerize copolymers via ring-opening and atom transfer radical polymerization is by using bifunctional initiators [35, 36]. The usage of a bifunctional initiator guarantees, that all chains have a functional group attached after the first polymerization step and additionally, this has the potential to save a working step when synthesizing more than one macroinitiator.

3.1.2 Ring-Opening Polymerization

Coordination-insertion, anionic, cationic, and nucleophilic polymerization are the most frequently reported controlled ring-opening polymerization (ROP) of cyclic monomers in the literature [37, 38]. The coordination-insertion and nucleophilic polymerization are undoubtedly the most efficient and general methods reported so far for the ROP of lactones, with cationic and anionic polymerization being much less investigated. While coordination-insertion polymerization uses metal-alkoxides and related complexes as catalysts, the organocatalytic nucleophilic polymerization is a metal-free approach to ROP.

Metal-free catalysts, like enzymes and organocatalysts, are attracting growing interest, since being considered to be the more economical and environmentally friendly alternatives [39]. The resulting polymers are particularly suitable for biomedical applications, since there is no concern of contamination with metals. Furthermore, the polymerization is typically performed at room temperature. Enzymatic polymerizations normally are slow and result in fairly small polymer yields with broad dispersities. In contrast, the organocatalytic ROP developed by Nederberg et al. in 2001 employs amine-based catalysts and shows extremely fast kinetics resulting in well-defined polymers and very high monomer conversion levels [39]. Since this discovery a vast range of organocatalysts based on pyridine, phosphine, and N-heterocyclic carbenes have been reported in the literature [5]. The brief introduction given here will only focus on the use of guanidine, amidine, and supramolecular thiourea-amine complexes as organocatalysts [40–42].

3.1.2.1 Organocatalytic ROP

The organocatalyst for this metal-free approach to ROP of cyclic monomers is a nucleophilic transesterification catalysts. The linear correlation between the molecular weight and conversion is proof for the living character of this nucleophilic

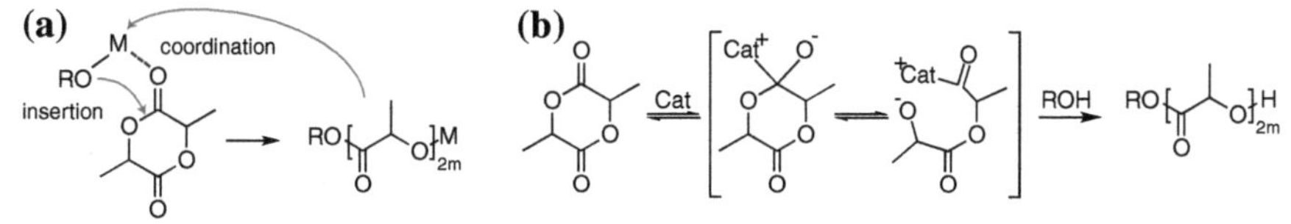

Fig. 3.7 Proposed mechanisms for lactide ROP techniques. **a** Once bound to the initiator the metal catalyst stays bonded to the growing polymer chain for the entire metal-catalyzed coordination-insertion ROP (Adapted with permission from Kamber et al. [5]. Copyright 2013 American Chemical Society). **b** For the nucleophilic ROP the organocatalyst, denoted Nu, activates the monomer that is added to the initiating protic agent or to the secondary alcohol group of the growing polymer chain, referred to as ROH (Adapted with permission from Dechy-Cabaret et al. [38]. Copyright 2013 American Chemical Society)

polymerization. Consequently, the degree of polymerization closely tracks the initial monomer/initiator ratio and polymers of controlled molecular weights and narrow polydispersities are obtained. The initiation of the ring-opening reaction requires the presence of a protic agent, typically water or an alcohol.

The proposed polymerization pathway differs fundamentally from the coordination-insertion mechanism involving metal complexes, see Fig. 3.7 [5, 38]. Indeed, the nucleophilic catalyst only activates the monomer toward ring opening, whereas the metal complex activates the monomer, initiates the polymerization, and remains bound to the growing chain. The polymerization mechanism of a superbase or thiourea-amine catalyzed ROP will be discussed in more detail below.

3.1.2.2 Components of Organocatalytic ROP

Monomers Figure 3.8 shows a small selection of cyclic monomers suitable for ROP [43]. Additionally, three different stereoisomers of lactide exist as a consequence of the presence of two stereocenters per monomer unit, namely *meso*-, L- and D-lactide, see Fig. 3.9. Further, racemic mixture of L- and D-lactide are commercially available. While ROP of either pure L- and D-lactide enables synthesis of highly crystalline poly(L-lactic acid) or poly(D-lactic acid), ROP of *rac*- or *meso*-lactide with adequate catalysts allows the synthesis of stereoblock copolymers, heterotactic and syndiotactic poly(lactic acid). Notably, stereoregular PLAs display much lower rates of degradation than the amorphous atactic polymer.

Initiators Primary and secondary alcohols as well as water were found to be efficient initiators, leading to the corresponding ester functionality at the growing

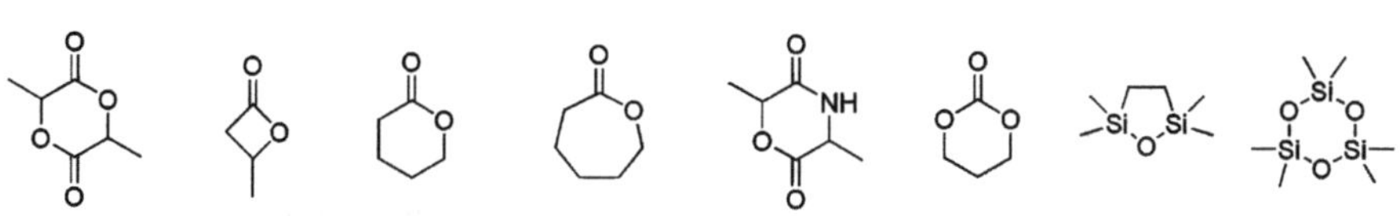

Fig. 3.8 Selection of cyclic monomers suitable for ROP (Adapted with permission from Dechy-Cabaret et al. [38]. Copyright 2013 American Chemical Society)

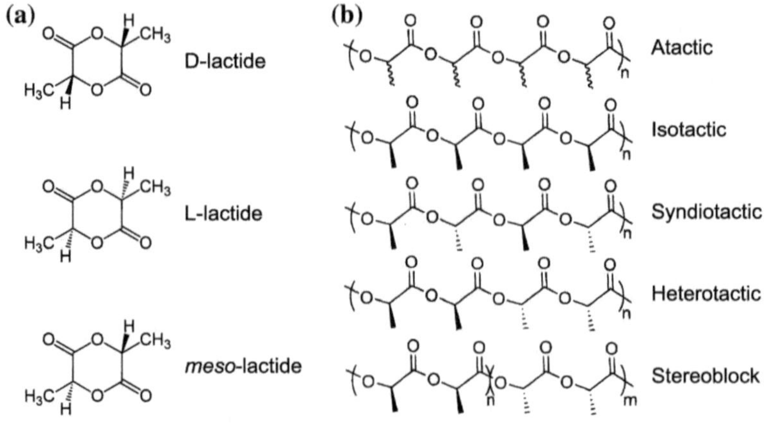

Fig. 3.9 Stereoisomerism of lactide and poly(lactic acid). **a** The three lactide stereoisomers D-lactide, L-lactide, and *meso*-lactide. **b** Stereosequences of poly(lactic acid) (Reproduced from Dove et al. [44] with permission of The Royal Society of Chemistry)

poly(lactic acid) chain end. In addition, for thiourea and tertiary amine based catalytic ROPs several protic functional groups were found to serve as initiators for lactide polymerization, including thiols, and silanols. Monohydroxy-functional macroinitiators were also used to initiate the synthesis of copolymers [5, 38].

Organocatalysts As mentioned earlier, organocatalytic ROP was first reported applying pyridines as the nucleophilic catalyst. Subsequently, guanidine, amidine, and thiourea-based organocatalysts were found to be highly active for the ROP of lactones [40–42].

Hedrick et al. reported the catalytic activity of a conjoined thiourea-tertiary amine molecule, displayed in Fig. 3.10a, in the ROP of lactide [42]. Mechanistic and theoretical studies support a supramolecular bifunctional mechanism involving activation of both the monomer and the alcohol nucleophile, see Fig. 3.10d. With reaction times

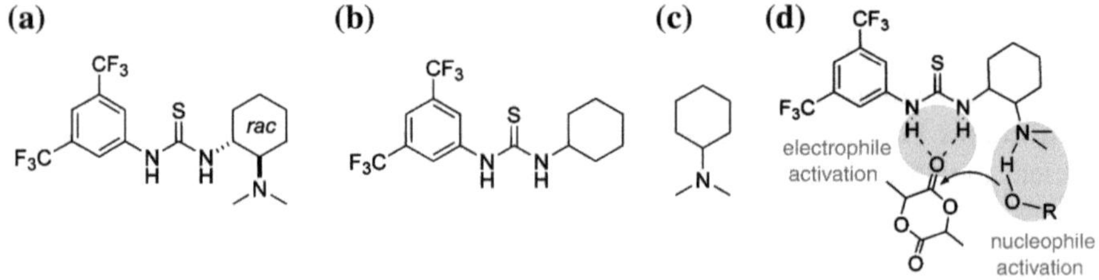

Fig. 3.10 Thiourea-based organocatalysts. **a–c** Chemical structure of the bifunctional thiourea-tertiary amine catalyst, bis(3,5-trifluoromethyl)phenyl cyclohexylthiourea (thiourea), and N,N-dimethylcyclohexylamine (Modified from Dove et al. [42]). **d** Proposed dual activation pathway of lactide ROP [41] (Adapted with permission from Pratt et al. [41]. Copyright 2013 American Chemical Society)

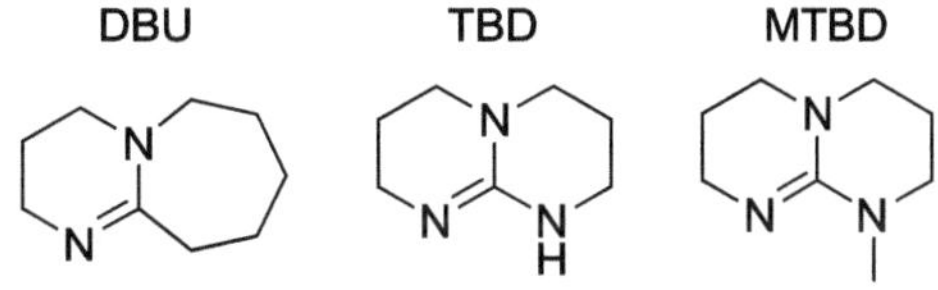

Fig. 3.11 Organocatalytic superbases. **a–c** Chemical structure of *DBU*, *TBD*, and *MTBD* [5]

of a few days this polymerization is significantly slower than that found for nitrogen-heterocyclic carbene catalysts. However, prolonged reaction times at ambient temperatures with the thiourea-based catalyst led to negligible polydispersity broadening even at near-complete monomer consumption, indicating that little transesterification of the linear polymer occurs, which might be of importance when synthesizing well-defined copolymers. Additionally, it was found that bis(3,5-trifluoromethyl)phenyl cyclohexylthiourea, hereafter referred to as thiourea, together with a tertiary amine (Fig. 3.10b, c) catalyze the ROP of lactones, while separately they are not capable of doing so [40–42].

Since the requirement for a highly electron withdrawing thiourea were given, it was proposed, that in order to increase the catalytic activity it is necessary to increase the basicity of the tertiary amine. Screening studies demonstrated that a variety of thiourea-amine combinations are catalytically active for the ROP of lactide [41]. The steep acceleration of polymerization attributed to the higher basicity of (−)-sparteine compared to the tertiary amine inspired Lohmeijer et al. to use even stronger organic bases, so-called superbases, such as the amidine 1,8-diazabicycloundec-7-ene (DBU), guanidine 1,4,7-triazabicyclodecene (TBD), and N-methylated TBD (MTBD), displayed in Fig. 3.11. Interestingly, these bases proved to be highly efficient catalysts for the ROP of lactide in nonpolar solvents without any thiourea [40]. In contrast to the thiourea-amine systems, their basicities are such that the thiourea is unnecessary for lactide polymerization, and complete conversions are reached extremely fast.

The enhanced activity of TBD, which is the most active of the three bases, was demonstrated by a model reaction in which a loading of 0.1 $\%_{mol}$ TBD to monomer was shown to polymerize 500 equivalents of lactide to 95 % monomer conversion within 1 min at ambient temperatures to produce a polymer with $M_n = 62.6\,\mathrm{kg\,mol^{-1}}$ and $PDI = 1.11$ [40]. However, a polydispersity broadening was observed at high conversions, making it difficult to synthesize well-defined copolymers when relying on a nearly complete monomer conversion to control the degree of polymerization.

The less active bases MTBD and DBU were also found to polymerize lactide reasonable fast, but with the advantage that after monomer consumption, hardly any transesterification occurred. The reason for the reduced activity of these bases compared to TBD are their structural differences, while TBD contains two accessible nitrogen atoms as illustrated in Fig. 3.12a, MTBD and DBU are essentially monofunctional.

TBD alone demonstrated remarkable activities for the ROP of δ-valerolactone and ε-caprolactone, while both DBU and MTBD can only polymerize these monomers in the presence of the thiourea co-catalyst, due to the discussed monofunctionality.

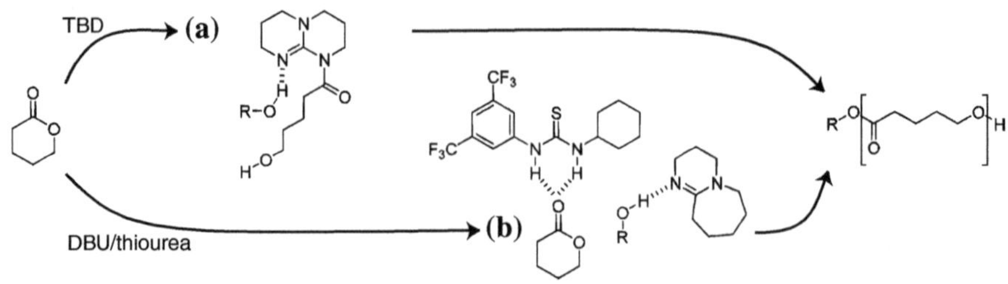

Fig. 3.12 ROP of valerolactone and caprolactone. **a** ROP of caprolactone catalyzed with the dual activation by TBD. **b** Vaprolactone ROP through the dual activation by DBU/thiourea (Adapted with permission from Lohmeijer et al. [40]. Copyright 2013 American Chemical Society)

The proposed pathway of ROP of valerolactone through the dual activation by DBU and thiourea is illustrated in Fig. 3.12b. To the best of the author's knowledge, no DBU or DBU/thiourea catalyzed ROP of lactide initiated by a macroinitiator has been reported.

Additives An advantage of organocatalytic ROP of lactide is that the synthesis is not oxygen sensitive. However, the reagents and solvents have to be extremely dry, since every water molecule initiates and deactivates/reacts with the catalysts, e.g. the strong bases. The initiation of poly(lactic acid) homopolymer by water is especially unwanted when polymerizing well-defined copolymers using a macroinitiator. Acids such as benzoic acid are used for a rapid termination of the polymerization. Chloroform and dichloromethane are commonly used as solvents, since lactide is easily dissolved in these solvents. Chloroform stabilized with amylenes rather than with ethanol should be used.

3.1.2.3 Organocatalytic ROP as Copolymerization

The beauty of using the highly efficient and fast organocatalysts for polymerization of the final lactide block of copolymers with well-defined block volume ratios is that almost complete monomer conversion is achieved. This fact enables to determine the desired copolymer composition by using calculated amounts of macroinitiator and lactide monomer. A convenient process to produce a series of copolymers with different volume ratios involves the preparation of stock solutions of the monomer and macroinitiator which are then mixed using an accurate micropipette. This is of course only possible when the reaction solution is free from other initiators like water and alcohols. For larger quantities of copolymer the appropriate amount of reagents can be added by weight.

Pratt et al. reported the synthesis of a series of lactide containing block copolymers using monohydroxy-functional macroinitiators and a thiourea/(−)-sparteine catalyst [41]. In another publication from the same group around Hedrick the synthesis of lactide, valerolactone, and caprolactone from monohydroxy-functional PS-OH, PEO-OH and PDMA-OH using TBD as catalyst was demonstrated [40].

3.2 Experimental Methods

3.2.1 Materials

The chemicals listed in Table 3.1 were used as received unless stated otherwise.

3.2.2 Characterization Methods

Characterization of molecular weight and polydispersity was performed by means of 500 MHz ^{1}H nuclear magnetic resonance (NMR) spectroscopy and size exclusion chromatography (SEC). The NMR proton chemical shifts of the PLA's methyl group, the PXS's backbone chain, and water overlap in the region of 1–2 ppm and hence do not contain easily accessible information about the copolymer composition. Fortunately, the peaks arising from the protons of the styrenic phenyl groups and the remaining CH group of the PLA backbone are well resolved in the region of 5–7.2 ppm and do not overlap with other peaks. Knowing the number of protons per monomer repeat unit of each block allows calculation of the ratio between the degree of polymerization of the blocks N_{PXS}:N_{PLA} by integrating the area under the corresponding peaks. From this value the weight ratio and the volume fraction of the two copolymer blocks can be derived using the molecular weight of the monomers and the density of the corresponding homopolymers found in the literature, respectively.

The SEC was equipped with a refractive index and photodiode array detector which were calibrated with poly(styrene) standards. Tetrahydrofuran was used as eluent and the data recorded by the photodiode array detector was evaluated at a wavelength of $\lambda = 257$ nm, where poly(styrene) has an absorption maximum. The poly(styrene)-equivalent molecular weight of the copolymers determined by the SEC were found to be higher than the theoretical values. More accurate values were calculated using the SEC determined molecular weight of the styrenic macroinitiator together with the weight ratio of the two copolymer chains extracted from NMR data.

3.2.3 Synthesis of *β*-hydroxylethyl *α*-bromoisobutyrate

Considering the activity data of the different ATRP initiators presented in Fig. 3.4, using β-hydroxylethyl α-bromoisobutyrate (HEBIB) as bifunctional initiator seemed reasonable and was synthesized according to White et al. [45]. Briefly, anhydrous ethylene glycol (90 ml, 1.6 mol) was cooled to 0 °C in a flame-dried two-neck round bottom flask. α-Bromoisobutyryl bromide (8 ml, 0.06 mol) was slowly added with an addition funnel to the stirring ethylene glycol. The reaction solution was stirred at 0 °C for three hours. Thereafter, the reaction was quenched with 40 ml of deionized water and extracted with chloroform (3 × 40 ml). The chloroform/product mixture

Table 3.1 Chemicals used for the synthesis of diblock copolymers

Chemical	Purity [%][a]	Abbreviation
1,1,4,7,10,10-Hexamethyltriethylenetetramine	97 (SA)	HMTETA
1,8-Diazabicyclo[5.4.0]undec-7-ene	98 (SA)	DBU
2,2′-Bipyridyl	+99 (SA)	Bipy
3,5-Bis(trifluoromethyl)phenyl isothiocyanate	+99 (AO)	–
3,6-Dimethyl-1,4-dioxane-2,5-dione	99 (AO)	Lactide
4-Bromostyrene	98 (SA)	4-BrS
4-Chlorostyrene	98 (SA)	4-ClS
4-Fluorostyrene	98 (AS)	4-FS
α-Bromoisobutyryl bromide	98 (SA)	–
Chloroform, extra dry, stabilized with amylene	99.9 (AO)	TCM
Copper(I) bromide	99.999 (SA)	–
Cyclohexylamine	+99 (AA)	–
Diphenyl ether	99 (SA)	PhOPh
Ethyl α-bromoisobutyrate	98 (SA)	EBIB
Ethylene glycol, anhydrous	99.8 (SA)	–
N,N,N',N",N"-Pentamethyldiethylenetriamine	99 (SA)	PMDETA
Phosphorus pentoxide, powder	98 (SA)	–
Styrene	+99 (SA)	–
Tin(II) 2-ethylhexanoate	95 (SA)	$Sn(EH)_2$
Tris[2-(dimethylamino)ethyl]amine	– (SA)	Me_6TREN

[a]Suppliers are given in brackets: Alfa aesar (AA), Apollo scientific (AS), Acros organics (AO), Sigma aldrich (SA)

was purified by washing with deionized water (3 × 100 ml). Instead of distilling the product, as suggested in [45], the chloroform/product mixture was purified by washing with deionized water (3 × 100 ml). Then the combined organic extracts were dried over $MgSO_4$, filtered, and the $CHCl_3$ was removed by a rotary evaporator. The NMR spectrum is shown in Fig. A.1.

3.2.4 Synthesis of bis(3,5-trifluoromethyl)phenyl Cyclohexylthiourea

The superbase DBU was chosen as catalyst for the ROP of lactide because of its fast kinetics, high efficiency, and prevention of transesterification. However, it was decided to add bis(3,5-trifluoromethyl)phenyl cyclohexylthiourea (thiourea) as co-catalyst to further increase the efficiency and overcome possible steric hindrance arising from the bulkiness of macromolecular initiators. The thiourea was synthesized according to Pratt et al. [41]. Briefly, 3,5-bis(trifluoromethyl)phenyl isothiocyanate (3.37 ml, 18.5 mmol) and anhydrous tetrahydrofuran (20 ml) were added to a flame-dried two-neck round bottom flask. Cyclohexylamine (2.11 ml, 18.5 mmol) was added dropwise via a syringe at room temperature to the stirring solution. After

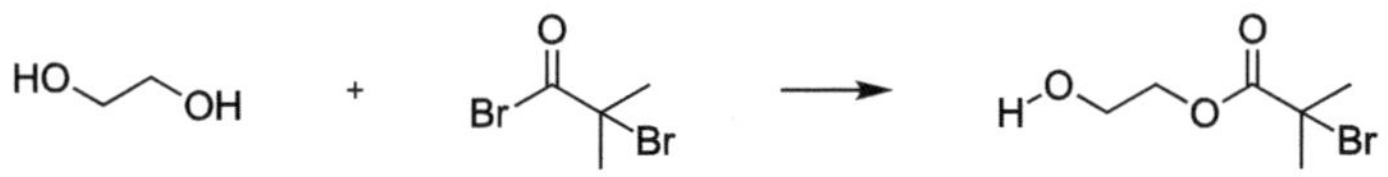

Fig. 3.13 Synthesis of the bifunctional initiator HEBIB

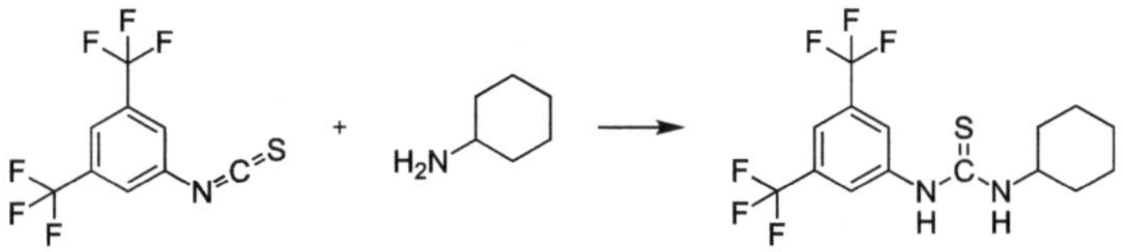

Fig. 3.14 Synthesis of the thiourea co-catalyst

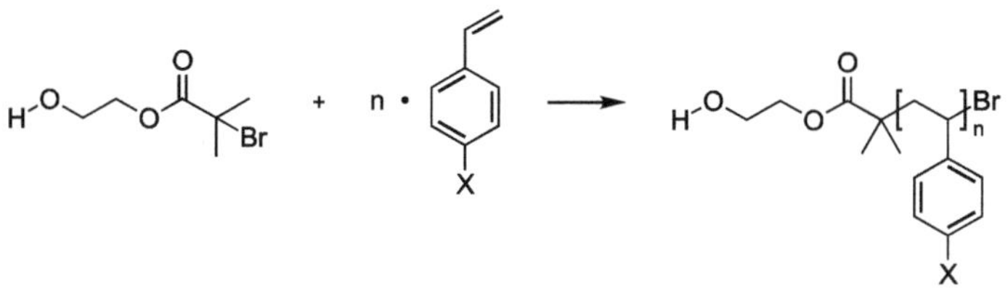

Fig. 3.15 ATRP of monohydroxy-functional styrenic polymers initiated by HEBIB

the solution stirred for four hours, the solvent was evaporated and the white residue was recrystallized three times from chloroform to give the thiourea as a white powder, which was dried under vacuum at 80 °C for 24 h. The NMR spectrum is shown in Fig. A.2.

3.2.5 General Procedure of Styrene ATRP

Typically, the styrenic monomer was run over a column of activated basic alumina to remove the inhibitor and collected into a flame-dried Schlenk tube, which was equipped with a magnetic stir bar. Polymerizations were performed in bulk or in a solvent, in the latter case the solvent was added to the reaction tube together with the monomer. Before and after adding the remaining reagents the reaction solution was deoxygenated by performing three freeze-pump-thaw cycles. The reagents, including CuBr, reducing agent, ligand, and initiator, were added to the degassed monomer solution under a constant flow of inert gas, with the initiator being the last reagent added. Then the tube was refilled with inert gas and placed in a preheated oil bath. At this point the CuBr dissolved and the reaction turned slightly green (Figs. 3.13, 3.14).

Small samples were taken from the reaction mixture during polymerization using a degassed syringe and analyzed by SEC. The polymerization rate and the appropriate reaction time were calculated from the molecular weight information

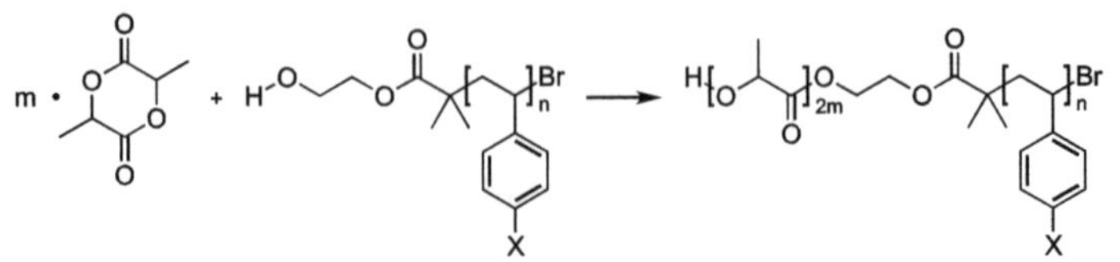

Fig. 3.16 ROP of lactide initiated by monohydroxy-functional styrenic polymer

obtained from the SEC measurements. Once the reaction time elapsed, the reaction was terminated by quenching the greenish-brown reaction mixture to room temperature. Subsequently, the polymer was dissolved in toluene or THF, run over a column of neutral alumina and silica, precipitated into a large excess of cold methanol, and collected by filtration. This purification process was repeated until the polymer was totally white indicating that most of the copper salt had been removed. The polymer was dried under vacuum at 80 °C for 24 h before characterization and further usage.

3.2.6 General Procedure of Organocatalytic Lactide ROP

The ROP of lactide was performed in a nitrogen glove box in order to avoid any traces of moisture contamination. Before transferring the 3,6-dimethyl-1,4-dioxane-2,5-dione (D,L-lactide) and the thiourea co-catalyst into the glove box they were either dried together with hydroscopic phosphorus pentoxide powder in a desiccator under reduced pressure for one week or in a vacuum oven at 60 °C for 24 h. The liquid catalyst, 1,8-diazabicyclo[5.4.0]undec-7-ene (DBU), was dried by stirring over CaH_2 for 24 h and filtered with nylon syringe filters (0.45 µm) in a glove box to remove the drying agent. As reaction vessel for the ROP glass vials equipped with a stir bar were used, which were heated up with a heat gun directly before putting them into the glove box's ante-chamber.

Adequate amounts of the (macro)initiator, monomer, and thiourea were added to the vial and dissolved in anhydrous dichloromethane (DCM) or amylene stabilized chloroform (TCM). Typically, the mass concentration of the solid reagents, i.e. lactide monomer and macroinitiator, was around 10 %. After stirring the solutions for a couple of minutes the DBU was added with an accurate micropipette. The vials were sealed and the reaction solution was stirred at room temperature. Once the reaction time had elapsed the DBU was neutralized by adding equimolar amounts of benzoic acid. Then the polymer was precipitated into cold methanol, collected by filtration, and dried under vacuum at 80 °C for 24 h. To reduce the risk of poly(lactic acid) degradation in the protic methanol, fast filtration is essential. This was achieved by allowing the precipitate to sediment and removing the excess methanol before filtration.

3.3 Results and Discussion

3.3.1 Test Polymerizations

Reasonable reaction parameters for the ATRP were determined first by using styrene and the commercially available ethyl α-bromoisobutyrate (EBIB) followed by small-scale reactions with 4-fluorostyrene (<1 g), before polymerizing larger quantities of expensive halogenated styrenes with the synthesized initiator HEBIB. Similarly, suitable reaction conditions for the ROP of lactide were tested using 1-pentanol as initiator instead of the synthesized macroinitiators.

3.3.1.1 ATRP Test Reactions

The first few reaction did not yield any polymer, most probably due to catalyst oxidation. After adding reducing agents, namely zero-valent copper or $Sn(EH)_2$, controlled polymerizations were observed that were less susceptible to oxygen contamination. A selection of the tested ATRP conditions together with the molecular weight data of the resulting polymers is presented in Table 3.2. High molecular weight polymers were intentionally synthesized in this early stage of the project with the aim to gain double-gyroid templates with large unit cell dimensions. It was however impossible to induce a morphology transition to the double-gyroid phase by temperature annealing for the high molecular weight copolymers, see Chap. 4.

Diphenyl ether (PhOPh) and anisole were probed as solvents, but no significant improvement in the PDI resulted. Lowering the reaction temperature from 120 to 80 °C decreased the reaction kinetics, but also increased the polydispersity. PMDETA was found to be the most suitable of the four tested ligands, especially in combination with Cu and CuBr, enabling good control over the polymerization kinetics. Sample PS15 exhibits the lowest PDI, although it has a comparably high molecular weight of 65 kg mol^{-1}. $Sn(EH)_2$ as reducing agent is too powerful, resulting in fast but uncontrolled ATRP as sample PFS7 demonstrates.

Bulk polymerizations with equimolar concentrations of CuBr, Cu, and PMDETA at 80 °C was identified as the most suitable for the synthesis of the hydroxy-functional styrenic macroinitiators.

3.3.1.2 ROP Test Reactions

In a first batch of lactide ROPs an appropriate solvent and catalyst were determined, see Table 3.3. The polydispersity of poly(lactic acid) synthesized in TCM compared to DCM was significantly lower. The addition of thiourea co-catalyst lead to an improved PDI in case of DCM, while no change was observed in TCM. However, when using poly(4-X-styrene)-OH as macroinitiator only the DBU/thiourea co-catalyst system was found to yield copolymer. Considering Fig. 3.12b, the lactide

Table 3.2 Molecular weight and PDI for different ATRP conditions

Sample	Reagents [mole ratio]	T [°C]	t [h]	$M_n \left[\frac{kg}{mol}\right]$	PDI
PS5	4-HS:EBIB:Cu:Me_6TREN:PhOPh 960:1:1:1:1400	110	16	58	1.41
PS8	4-HS:EBIB:Cu:Bipy:PhOPh 750:1:2:6:1100	110	16	22	1.32
PS10	4-HS:EBIB:Cu:Bipy:PhOPh 750:1:3:9:1000	110	20	54	1.38
PS11	4-HS:EBIB:CuBr:Cu:Me_6TREN 750:1:1:1:1	90	16	51	1.44
PS12	4-HS:EBIB:CuBr:Cu:Bipy 750:1:1:1:3	90	20	16	1.59
PS13	4-HS:EBIB:Cu:Bipy 960:1:2:6	90	15	63	1.42
PS14	4-HS:EBIB:Cu:PMDETA 960:1:2:2	90	15	33	1.30
PS15	4-HS:EBIB:CuBr:Cu:PMDETA 960:1:1:1:1	90	15	65	1.21
PS16	4-HS:EBIB:CuBr:Cu:HMTETA 960:1:1:1:1	80	20	68	1.29
PS17	4-HS:EBIB:CuBr:$Sn(EH)_2$:HMTETA 960:1:1:0.5:1	80	20	83	1.32
PS19	4-HS:EBIB:CuBr:Cu:HMTETA 960:1:1:1:1	120	20	70	1.39
PS20	4-HS:EBIB:CuBr:Cu:PMDETA 960:1:1:1:1	120	20	12	1.37
PS21	4-HS:EBIB:CuBr:Cu:PMDETA:Anisole 960:1:1:1:1:1000	120	20	38	1.53
PFS4	4-FS:EBIB:CuBr:Cu:PMDETA 700:1:1:1:1	80	14	39	1.13
PFS7	4-FS:HEBIB:CuBr:$Sn(EH)_2$:PMDETA 750:1:1:0.5:1	80	16	122	1.57
PFS8	4-FS:HEBIB:CuBr:Cu:HMTETA 750:1:1:1:1	80	16	59	1.18
PFS13	4-FS:HEBIB:CuBr:Cu:PMDETA 800:1:1.1:1.1:1.1	80	18.5	53	1.12
PFS14	4-FS:HEBIB:CuBr:Cu:PMDETA 800:1:1.1:1.1:2.2	80	20	62	1.16
PFS15	4-FS:HEBIB:CuBr:Cu:PMDETA 800:1:1.1:1.1:1.1	80	24	72	1.15
PFS20	4-FS:HEBIB:CuBr:Cu:PMDETA:Anisole 570:1:1.1:1.1:1.1:570	80	19	24.5	1.14
PFS21	4-FS:HEBIB:CuBr:PMDETA 330:1:3:3	90	4.5	36	1.14

Table 3.3 Effects of ROP conditions on the molecular weight of poly(lactic acid)

Sample	Reagents [mole ratio][a]	t [min]	$M_n \left[\frac{kg}{mol}\right]^b$	PDI
PLA1	Lactide:pentanol:DBU:DCM 200:1:2:4400	140	23	1.37
PLA2	Lactide:pentanol:DBU:TCM 200:1:2:3300	140	25	1.11
PLA3	Lactide:pentanol:DBU:thiourea:DCM 200:1:0.5:2:4400	140	20	1.22
PLA4	Lactide:pentanol:DBU:thiourea:TCM 200:1:0.5:2:3300	140	20	1.11

[a] Approximate reagent ratio
[b] Poly(styrene)-equivalent molecular weight

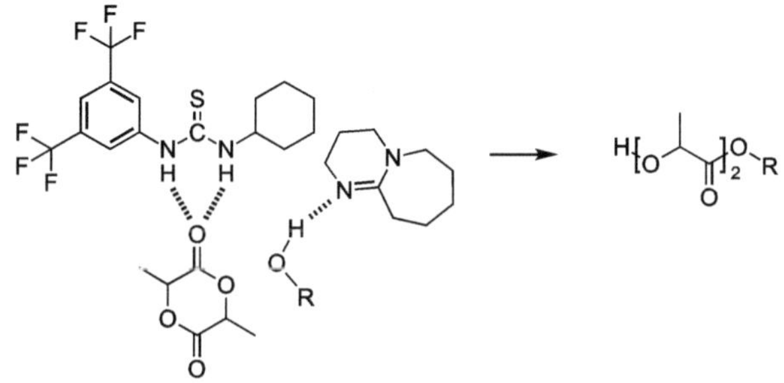

Fig. 3.17 Proposed lactide ROP mechanism through the dual activation of DBU/thiourea

polymerization mechanism based on a dual activation by DBU and the thiourea shown in Fig. 3.17 is proposed.

To test whether the DBU/thiourea catalyst leads to transesterification during prolonged reaction times a sequential series of samples were taken after different reaction times from a ROP and quenched with benzoic acid. The reaction solution contained 1-pentanol as initiator and an equimolar amount of DBU and thiourea as catalyst. The molar ratio of lactide:catalyst was 50:1 and the monomer mass concentration in chloroform was 10 %. The data presented in Table 3.4 suggests, that the catalyst system is highly efficient with an almost complete monomer conversion within the first 45 min and that no transesterification occurs during prolonged reaction times.

Table 3.4 Effect of reaction time on the polydispersity of poly(lactic acid)

Sample	t [min]	$M_n \left[\frac{kg}{mol}\right]^a$	PDI
PLA5_1	15	9.2	1.09
PLA5_2	45	15.2	1.07
PLA5_3	90	15.7	1.09

Table 3.5 Polymerization conditions and molecular weight of P(F)S-OH containing varying amounts of fluorinated monomer and PXS-OH

Sample	4-FS [%][a]	m_M [g][b]	M:I:C [mole ratio][c]	t [h]	M_n $\left[\frac{kg}{mol}\right]$	PDI
PFS17	100	9	570:1:1.1	7	50	1.18
PFS18	100	8	250:1:0.8	3.6	15	1.13
PFS19	100	7	270:1:1.1	4	30	1.17
PFS23	100	5	370:1:0.8	5.5	24	1.13
PFS24	100	6	300:1:0.8	5.2	21	1.12
PFS26	100	24	200:1:0.7	7	17	1.10
PFS27	100	25	250:1:0.7	8	33	1.09
PFS28	100	25	300:1:0.7	8	36	1.10
PFS29	100	10	120:1:0.7	8	13	1.07
PFS37	100	25	200:1:0.7	7.3	20	1.07
PFS38	100	10	210:1:0.7	6	19	1.06
P(F)S39	95	13	200:1:0.7	5.2	19	1.08
P(F)S40	90	13	200:1:0.7	5.2	17	1.08
P(F)S41	80	13	200:1:0.7	5.2	18	1.08
P(F)S42	60	16	200:1:0.7	5.5	17	1.09
PFS43	100	20	190:1:0.7	7.7	19	1.08
PFS44	100	17	180:1:0.6	8	15	1.22
PFS45	100	8	150:1:0.6	5	15	1.17
P(F)S46	20	16	200:1:0.6	5.5	18	1.15
PS47	0	30	210:1:0.6	6	17	1.15
PFS48	100	8	120:1:0.5	4	11	1.13
P(F)S49	30	15	150:1:0.5	4.5	14	1.12
PS50	0	20	150:1:0.5	5	13	1.12
PBrS1	–	1	100:1:0.5	3.5	12	1.43
PBrS2	–	1	100:1:0.5	3.5	13	1.09
PBrS3	–	19	100:1:0.5	3.5	11	1.07
PClS1	–	1	110:1:0.5	3.5	13	1.09
PClS2	–	1	110:1:0.5	3.5	12	1.08
PClS3	–	10	110:1:0.5	3.5	11	1.06

[a]4-Fluorostyrene fraction of the initial monomer loading
[b]Weight of initial monomer (M) loading
[c]HEBIB was used as initiator (I) and equimolar concentration of CuBr:Cu:PMDETA as catalyst (C)

3.3.2 *Synthesis of Poly(4-X-styrene)-OH via ATRP Initiated by HEBIB*

A range of monohydroxy-functional poly(4-X-styrene) (PXS) macroinitiators for the subsequent conversion into diblock copolymers via ROP of lactide were synthesized according to the general procedure described in Sect. 3.2.5 with the optimized reaction conditions as determined by the test reactions, namely bulk polymerization with equimolar concentrations of CuBr, Cu, and PMDETA at 80 °C initiated by HEBIB.

The synthesis was fairly reproducible, yielding polymers with polydispersities generally below 1.22, see Table 3.5.

First, PFS-OH macroinitiators were prepared which only contained 4-fluorostyrene similar to the styrenic block of Prof. Hillmyer's double-gyroid-forming copolymer MHI41 which was prepared by metal-catalyzed ROP and RAFT polymerization. Using the same monomers as used for MHI41 allowed to test if the new two-step synthesis route employing ATRP and organocatalytic ROP is suited to synthesize similar double-gyroid-forming copolymers. After obtaining positive results with the newly synthesized PFS-*b*-PLA, see Chap. 4, monomer mixtures of 4-fluorostyrene and styrene were used to synthesize P(F)S-OH macroinitiators. The fraction of the expensive fluorinated monomer was gradually reduced from 100 to 0 % without having any notable effect on the polydispersity or other polymerization parameters, see Table 3.5. Since styrene and 4-fluorostyrene show similar ATRP kinetics, as discussed in 3.1.1.4, polymerization of their monomer mixtures are expected to yield random copolymers poly(4-fluorostyrene-*r*-styrene). Depending on the amount of fluorinated monomer the macroinitiators are denoted as PFS, P(F)S, and PS in the following. Furthermore, 4-bromostyrene and 4-chlorostyrene were used to polymerize ultraviolet cross-linkable PBrS and PClS, respectively.

During the course of the different polymerizations the catalyst loading was gradually reduced from a HEBIB:catalyst molar ratio of 1:1.1 to 1:0.5. This significantly reduced the necessary costly purification steps to remove the copper complexes after polymerization. To make this synthesis route towards double-gyroid-forming copolymers economically viable the catalyst loading would need to be further reduced to the ppm level. Although possible, this was not attempted in this study.

High monomer conversions were intentional in order to minimize the waste of expensive monomers, although this led to a steep viscosity increase of the reaction mixture and even solidification. To examine if this change in viscosity and the reaction conditions supported a living chain growth, samples were drawn from the reaction solution of PFS23 at different stages of the polymerization and analyzed by means

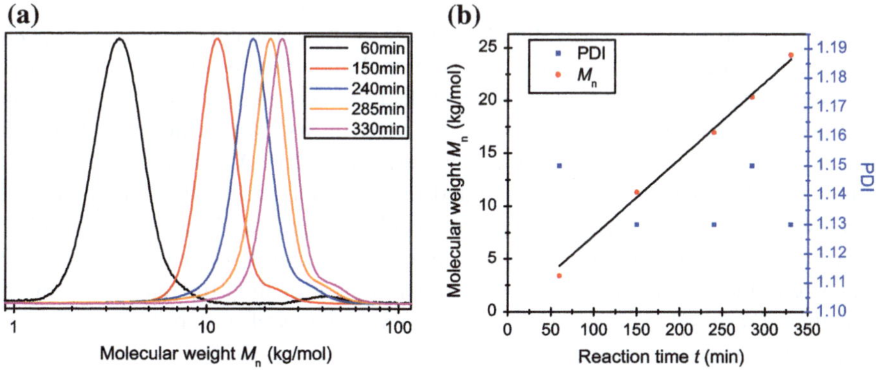

Fig. 3.18 **a** SEC traces of samples taken at different reaction times from the reaction solution of PFS23. **b** Evolution of molecular weight and polydispersity of PFS23 during the polymerization

of SEC. The SEC traces of five consecutive samples are shown in Fig. 3.18a. A high molecular weight shoulder is observed as the monomer consumption proceeded, this was most likely due to coupling reactions. However, similar shoulders were also present in all earlier performed test reactions, regardless of the employed polymerization conditions, and the shoulder formation has been reported in the literature [46]. A constant, linear average molar mass gain of 72 $kg\,mol^{-1}\,min^{-1}$ is a good indicator for a well controlled living polymerization, see Fig. 3.18b. A similar polymerization rate of 45.5 and 39.1 $kg\,mol^{-1}\,min^{-1}$ was found for P(F)S49 and PS50, respectively. Further, an increase in polydispersity was not observed even when a molecular weight of 24 $kg\,mol^{-1}$ was reached, which is close to the theoretical maximum of 25 $kg\,mol^{-1}$, calculated based on the initial monomer to initiator loading. When the reaction solidifies the living polymer ends are less likely to meet, thereby preventing coupling reactions to occur during prolonged reaction times. Calculating the chain growth rate by analyzing samples taken during the polymerization of the macroinitiators allows to abort the reaction when a desired molecular weight is reached. As determined in Chap. 4, PXS-OH with a molecular weight between 10 and 20 $kg\,mol^{-1}$ are suitable macroinitiators for double-gyroid-forming copolymers.

*3.3.3 Synthesis of Poly(4-X-styrene)-b-poly(*D,L*-lactic acid) via ROP Initiated by PXS-OH*

Considering the results obtained from the ROP test reactions in Sect. 3.3.1.2, the lactide block of poly(4-X-styrene)-*b*-poly(D,L-lactic acid) (PXS-*b*-PLA) was polymerized according to the general synthesis procedure described in Sect. 3.2.6 under the following conditions: Equimolar amounts of DBU and thiourea were chosen as catalyst and chloroform as solvent. The molar ratio of lactide:catalyst was 50:1 and the combined mass concentration of the monomer and macroinitiator in chloroform was 10 %. The stirring mixtures were allowed to react at room temperate for at least 80 min.

In addition to the expected PXS-*b*-PLA peak also a low molecular peak was also present in the SEC curves of the first batches of the prepared copolymers. This peak indicated the unwanted synthesis of PLA homopolymer initiated by water molecules. After repeating the drying procedures and replacing some reagents, the contamination with homopolymer was eliminated. While DBU on its own without the thiourea co-catalyst is capable to catalyze the polymerization of PLA homopolymer, it was found, that the copolymerization initiated by PXS-OH was hardly supported and only a smaller than expected increase in molecular weight was detected by SEC after a reaction time of 80 min. Further, when omitting the neutralization of the superbase with an equimolar amount of benzoic acid before precipitating the reaction solution in methanol resulted in copolymers which did not adopt well-ordered morphologies during microphase separation.

Once all reagents were anhydrous and the initial issues with the copolymerization were resolved, the next challenge was to identify the exact weight ratio of lactide monomer to PXS-OH macroinitiator that results in double-gyroid-forming diblock copolymers. More precisely, apart from a few exceptions, copolymers with a PLA volume fraction smaller than 50 % were synthesized targeting double-gyroids with a styrenic matrix phase. NMR and SEC data for a selection of the around 230 synthesized diblock copolymers is presented in Table 3.6, while the complete list is presented in Table B.1.

Initially, small-scale polymerization with typically less than 50 mg of macroinitiator were performed (PXS-*b*-PLA1 to PXS-*b*-PLA115). Preparing such small-scale reactions with accurate reagent weight ratios by weighing each reagent individually is nearly impossible. Instead, mixtures with different ratios of the macroinitiator and lactide were prepared by volume using a pipette. For that purpose, larger amounts of each reagent were dissolved in chloroform. Although a precise micropipette was used, the PLA volume fraction of total copolymer determined by NMR f_{PLA}^{NMR} randomly deviated from the theoretical value f_{PLA}^{theo}, see Table. 3.6. Starting with sample PXS-b-PLA116 larger amounts of the precursor materials, at least 500 mg of macroinitiator, were used. The scale located in the glove box was accurate enough to prepare accurate monomer to initiator ratios and thus, predissolving the reagents was omitted. For these samples f_{PLA}^{NMR} was consistently lower than f_{PLA}^{theo} by around 1 %, which corresponds to a monomer conversion of about 98–99 %.

As presented in Fig. 3.19a the addition of the PLA block to the PXS-OH only resulted in the expected shift to higher molecular weights without having any visible effect on the shape or width of the SEC trace. In fact, the copolymer polydispersity was typically lower than the PDI of the macroinitiators. Closely examining the NMR chemical shift of the PLA CH group synthesized by organocatalyst and comparing it to MHI41 PLA block prepared by trimethylaluminium-catalyzed coordination-insertion ROP reveals pronounced differences, see Fig. 3.19b. Although a racemic monomer mixture was used in both cases, different stereosequences of the poly(lactic acid), namely heterotactic and stereoblock PLA, can be caused by the particular catalyst used [44, 47]. It was feared that the resulting differences in polymer properties, such as different glass transition temperatures and chain crystallization, would negatively affect the microphase separation behavior [48]. Fortunately, the organocatalytically synthesized PXS-*b*-PLA also equilibrated in the double-gyroid nanostructure as indicated in Table 3.6.

As discussed in Sect. 3.2.2 the molecular composition of the synthesized copolymers was determined by ^{1}H NMR. The relevant spectral range from 5.1 to 7.2 ppm for PFS18-*b*-PLA76, P(F)S49-*b*-PLA204, and PS50-*b*-PLA216 is presented in Fig. 3.20, while the full range spectra are available in the Appendix A. Further, assuming the densities $\rho_{PLA} = 1.18\,g\,cm^{-3}$, $\rho_{PFS} = 1.17\,g\,cm^{-3}$, and $\rho_{PS} = 1.05\,g\,cm^{-3}$ the corresponding PLA volume fractions were calculated [49, 50]. These were density values were also used for the calculation of the theoretical volume fractions. Note that the value for ρ_{PLA} was determined for a temperature of 180 °C and is considerably lower than the value of $1.25\,g\,cm^{-3}$ at room temperature.

Importantly, once the double-gyroid-forming monomer to macroinitiator ratio had been determined for a specific composition of the styrenic block, it was fairly

Table 3.6 Composition, molecular weight, and PDI of selected PXS-*b*-PLAs determined by means of SEC and NMR

Sample	m_{PXS} [mg][a]	f_{PLA}^{theo} [%][b]	f_{PLA}^{NMR} [%][b]	M_n^{SEC} $\left[\frac{kg}{mol}\right]^c$	M_n^{NMR} $\left[\frac{kg}{mol}\right]^c$	N^d	PDI	Structure[e]
MHI41[f]	–	–	40.0	36	–	380	1.21	**DG**
MHI41[g]	–	–	39.2	28	–	–	1.35	**DG**
PFS17-*b*-PLA17	25	40.0	–	61	–	–	1.46	C
PFS17-*b*-PLA18	25	40.0	–	72	–	–	1.26	C
PFS19-*b*-PLA27	50	42.0	–	55	–	–	1.17	C
PFS18-*b*-PLA32	25	42.5	–	32	–	–	1.18	**DG**, C
PFS18-*b*-PLA33	25	45.0	–	29	–	–	1.19	**DG**, H
PFS19-*b*-PLA34	25	32.5	–	36	–	–	1.30	–
PFS19-*b*-PLA39	25	45.0	–	49	–	–	1.27	C, L , H
PFS18-*b*-PLA47	50	41.0	–	32	–	–	1.12	**DG**
PFS19-*b*-PLA50	50	41.0	–	56	–	–	1.17	C, W
PFS17-*b*-PLA53	50	41.0	–	83	–	–	1.19	C, W
PFS19-*b*-PLA55	50	41.0	–	58	–	–	1.18	C, L
PFS17-*b*-PLA56	50	41.0	–	82	–	–	1.18	–
PFS18-*b*-PLA76	50	43.5	41.0	32	26	269	1.09	**DG**, C, L
PFS19-*b*-PLA79	50	47.5	–	58	–	–	1.13	C
PFS19-*b*-PLA80	50	50.0	–	55	–	–	1.14	C
PFS18-*b*-PLA82	50	45.0	46.0	–	28	301	–	**DG**, C, H
PFS23-*b*-PLA87	40	40.5	38.7	44	39	408	1.08	C, H
PFS23-*b*-PLA91	40	46.5	–	49	–	–	1.18	C, H
PFS23-*b*-PLA94	40	51.0	48.9	53	47	517	1.18	L
PFS18-*b*-PLA97	35	41.3	38.7	–	25	256	–	**DG**, C
PFS23-*b*-PLA103	35	40.5	38.7	47	39	408	1.17	C, W
PFS26-*b*-PLA116	500	43.0	42.3	33	30	314	1.08	C, L
PFS26-*b*-PLA118	500	38.0	36.9	–	27	278	–	**DG**
PFS26-*b*-PLA119	500	39.0	37.9	–	27	285	–	**DG**
PFS26-*b*-PLA127	500	38.0	36.4	31	27	275	1.07	C
PFS27-*b*-PLA130	500	39.0	36.9	–	52	541	–	C
PFS26-*b*-PLA132	500	38.3	37.4	–	27	281	–	**DG**
PFS26-*b*-PLA133	500	38.5	37.1	–	27	280	–	**DG**
PFS26-*b*-PLA136	500	39.0	–	28	–	–	1.12	**DG**, C
PFS37-*b*-PLA148	500	38.5	37.3	35	32	330	1.07	C
P(F)S39-*b*-PLA150	500	38.5	37.6	–	31	318		**DG**, W
P(F)S49-*b*-PLA192	500	38.3	37.9	28	23	256	1.10	**DG**
P(F)S49-*b*-PLA204	500	38.3	–	28	–	–	1.10	**DG**
PS50-*b*-PLA214	1,000	37.5	37.1	26	22	244	1.09	**DG**

[a] Amount of macroinitiator PXS-OH used
[b] Theoretical and experimentally determined volume fraction of PLA
[c] Total molecular weight determined by NMR and poly(styrene) calibrated SEC
[d] Total degree of polymerization calculated using the NMR data
[e] Morphologies observed in microphase separated films thermally annealed under various conditions
[f] As determined by Marc Hillmyer
[g] As determined in this study

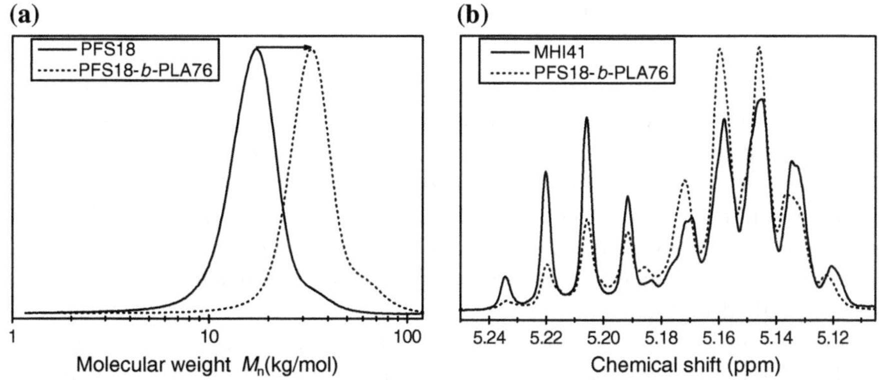

Fig. 3.19 **a** SEC traces of PFS18 and PFS18-*b*-PLA76. After addition of the PLA block the peak is shifted to higher molecular weights. **b** Differences in the ^{1}H NMR spectral range associated with to PLA's CH proton for MHI41 and PFS18-*b*-PLA76 in $CDCl_3$. For comparison, the area under the curves was normalized

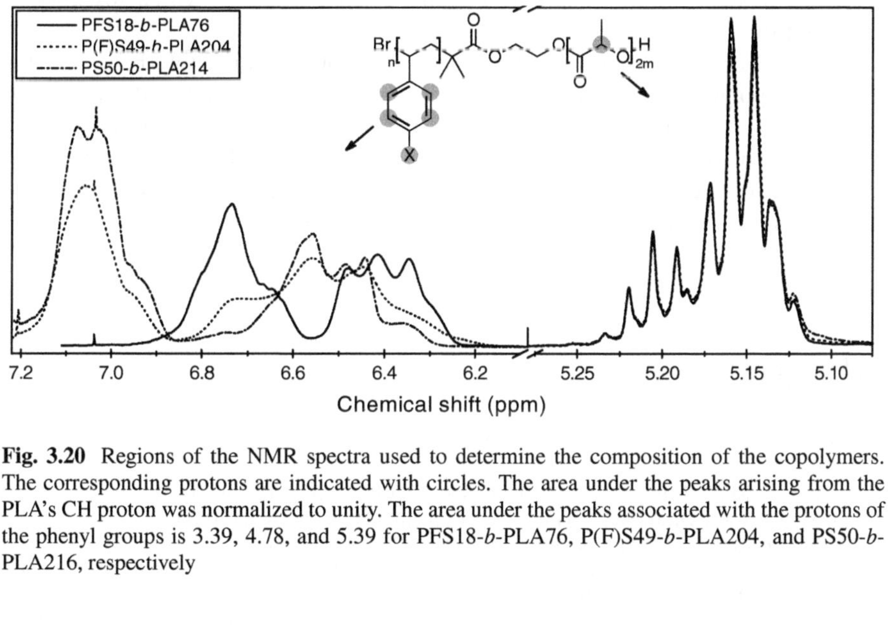

Fig. 3.20 Regions of the NMR spectra used to determine the composition of the copolymers. The corresponding protons are indicated with circles. The area under the peaks arising from the PLA's CH proton was normalized to unity. The area under the peaks associated with the protons of the phenyl groups is 3.39, 4.78, and 5.39 for PFS18-*b*-PLA76, P(F)S49-*b*-PLA204, and PS50-*b*-PLA216, respectively

unproblematic to synthesize the desired double-gyroid-forming diblock copolymers on a gram scale, see Table. 3.6 and B.1.

3.4 Concluding Remarks

In summary, almost 80 distinct ATRP polymerizations of 4-X-styrene monomers and over 215 copolymerization via lactide ROP were conducted. A set of optimized polymerization parameters were successfully identified that allowed the highly reproducible synthesis of well-defined diblock copolymers on a gram scale that adopt the double-gyroid nanomorphology during microphase separation as will be discussed in the following chapter. To reduce the synthesis cost and make the polymers more environmentally sustainable, the fraction of expensive fluorinated styrene monomer was gradually decreased.

However, to draw a final conclusion as to whether the presented synthesis route can deliver economically viable polymeric materials for templated-assisted manufacturing of nanomaterials, the following question has to be experimentally clarified: Can the involved ATRP and ROP be scaled up to yield larger batches of double-gyroid-forming material of at least 1–100 kg? This would however require reduction of the amount of copper catalyst for the ATRP to the ppm range. One kilogram of polymeric material would be sufficient to coat an area of $100\,m^2$ with a $10\,\mu m$ thick template film.

Further, I propose to attempt the synthesis of DG-forming poly(benzyl methacrylate)-*b*-poly(styrene)-*b*-poly(lactic acid) or poly(trimethylsilyl methacrylate)-*b*-poly(styrene)-*b*-poly(lactic acid) using HEBIB as initiator in the following order: First, ATRP of styrene; second, ATRP of the substituted methacrylate; and finally, ROP of lactide. Selective and successive degradation of the methacrylate and lactide blocks would enable to separately address and replicate the two double-gyroid networks.

References

1. M.R.J. Scherer, L. Li, P.M.S. Cunha, O.A. Scherman, U. Steiner, Adv. Mater. **24**(9), 1217–1221 (2012)
2. J. Rzayev, M.A. Hillmyer, J. Am. Chem. Soc. **127**(38), 13373–13379 (2005)
3. G. Moad, E. Rizzardo, S.H. Thang, Aust. J. Chem. **58**(6), 379–410 (2005)
4. K. Matyjaszewski, J. Xia, Chem. Rev. **101**(9), 2921–2990 (2001)
5. N.E. Kamber, W. Jeong, R.M. Waymouth, R.C. Pratt, B.G.G. Lohmeijer, J.L. Hedrick, Chem. Rev. **107**(12), 5813–5840 (2007)
6. A.J. Meuler, C.J. Ellison, J. Qin, C.M. Evans, M.A. Hillmyer, F.S. Bates, J. Chem. Phys. **130**(23), 234903–17 (2009)
7. S. Imamura, J. Electrochem. Soc. **126**(9), 1628–1630 (1979)
8. A. Andrady, *Physical Properties of Polymers Handbook*, 2nd edn. (Springer, Berlin Heidelberg, 2007), pp. 857–866
9. Q. Lin, *Physical Properties of Polymers Handbook* (Springer, Berlin Heidelberg, 2007), pp. 965–979
10. J.-S. Wang, K. Matyjaszewski, J. Am. Chem. Soc. **117**(20), 5614–5615 (1995)
11. J.-S. Wang, K. Matyjaszewski, Macromolecules **28**(23), 7901–7910 (1995)
12. M. Szwarc, Nature **178**(4543), 1168–1169 (1956)
13. T.E. Patten, J. Xia, T. Abernathy, K. Matyjaszewski, Science **272**(5263), 866–868 (1996)

14. V. Percec, T. Guliashvili, J.S. Ladislaw, A. Wistrand, A. Stjerndahl, M.J. Sienkowska, M.J. Monteiro, S. Sahoo, J. Am. Chem. Soc. **128**(43), 14156–14165 (2006)
15. W.A. Braunecker, K. Matyjaszewski, Prog. Polym. Sci. **32**(1), 93–146 (2007)
16. K. Matyjaszewski, N.V. Tsarevsky, Nat. Chem. **1**(4), 276–288 (2009)
17. T.E. Patten, K. Matyjaszewski, Adv. Mater. **10**(12), 901–915 (1998)
18. W. Jakubowski, K. Matyjaszewski, Macromolecules **38**(10), 4139–4146 (2005)
19. W. Jakubowski, K. Min, K. Matyjaszewski, Macromolecules **39**(1), 39–45 (2006)
20. W. Jakubowski, B. Kirci-Denizli, R.R. Gil, K. Matyjaszewski, Macromol. Chem. Phys. **209**(1), 32–39 (2008)
21. K. Matyjaszewski, T.E. Patten, J. Xia, J. Am. Chem. Soc. **119**(4), 674–680 (1997)
22. K. Matyjaszewski, W. Jakubowski, K. Min, W. Tang, J. Huang, W.A. Braunecker, N.V. Tsarevsky, PNAS **103**(42), 15309–15314 (2006)
23. K. Matyjaszewski, K. Davis, T.E. Patten, M. Wei, Tetrahedron **53**(45), 15321–15329 (1997)
24. V. Coessens, T. Pintauer, K. Matyjaszewski, Prog. Polym. Sci. **26**(3), 337–377 (2001)
25. T.E. Patten, K. Matyjaszewski, Acc. Chem. Res. **32**(10), 895–903 (1999)
26. J. Queffelec, S.G. Gaynor, K. Matyjaszewski, Macromolecules **33**(23), 8629–8639 (2000)
27. T. Pintauer, K. Matyjaszewski, Chem. Soc. Rev. **37**(6), 1087–1097 (2008)
28. J. Pietrasik, H. Dong, K. Matyjaszewski, Macromolecules **39**(19), 6384–6390 (2006)
29. K. Matyjaszewski, S. Coca, S.G. Gaynor, M. Wei, B.E. Woodworth, Macromolecules **31**(17), 5967–5969 (1998)
30. J. Qiu, K. Matyjaszewski, Macromolecules **30**(19), 5643–5648 (1997)
31. J. Xia, K. Matyjaszewski, Macromolecules **30**(25), 7697–7700 (1997)
32. K. Matyjaszewski, S. Coca, S.G. Gaynor, M. Wei, B.E. Woodworth, Macromolecules **30**(23), 7348–7350 (1997)
33. M.L. Becker, E.E. Remsen, K.L. Wooley, J. Polym. Sci., Part A. Polym. Chem. **39**(23), 4152–4166 (2001)
34. T.L. Bucholz, S.P. Li, Y.-L. Loo, J. Mater. Chem. **18**(5), 530–536 (2008)
35. A. Likhitsup, A. Parthiban, C.L.L. Chai, J. Polym. Sci., Part A. Polym. Chem. **46**(1), 102–116 (2008)
36. L. Tao, B. Luan, C-y Pan, Polymer **44**(4), 1013–1020 (2003)
37. C. Jérôme, P. Lecomte, Adv. Drug Delivery Rev. **60**(9), 1056–1076 (2008)
38. O. Dechy-Cabaret, B. Martin-Vaca, D. Bourissou, Chem. Rev. **104**(12), 6147–6176 (2004)
39. F. Nederberg, E.F. Connor, M. Möller, T. Glauser, J.L. Hedrick, Angew. Chem. Int. Ed. **40**(14), 2712–2715 (2001)
40. B.G.G. Lohmeijer, R.C. Pratt, F. Leibfarth, J.W. Logan, D.A. Long, A.P. Dove, F. Nederberg, J. Choi, C. Wade, R.M. Waymouth, J.L. Hedrick, Macromolecules **39**(25), 8574–8583 (2006)
41. R.C. Pratt, B.G.G. Lohmeijer, D.A. Long, P.N.P. Lundberg, A.P. Dove, H. Li, C.G. Wade, R.M. Waymouth, J.L. Hedrick, Macromolecules **39**(23), 7863–7871 (2006)
42. A.P. Dove, R.C. Pratt, B.G.G. Lohmeijer, R.M. Waymouth, J.L. Hedrick, J. Am. Chem. Soc. **127**(40), 13798–13799 (2005)
43. G. Rokicki, Prog. Polym. Sci. **25**(2), 259–342 (2000)
44. A.P. Dove, Chem. Commun. **48**, 6446–6470 (2008)
45. M.A. White, J.A. Johnson, J.T. Koberstein, N.J. Turro, J. Am. Chem. Soc. **128**(35), 11356–11357 (2006)
46. Y. Shen, S. Zhu, F. Zeng, R. Pelton, Macromol. Chem. Phys. **201**(13), 1387–1394 (2000)
47. K.A.M. Thakur, R.T. Kean, E.S. Hall, J.J. Kolstad, E.J. Munson, Macromolecules **31**(5), 1487–1494 (1998)
48. H. Uehara, Y. Karaki, S. Wada, T. Yamanobe, ACS Appl. Mater. Interfaces **2**(10), 2707–2710 (2010)
49. D.R. Witzke, R. Narayan, J.J. Kolstad, Macromolecules **30**(23), 7075–7085 (1997)
50. K.M. Nampoothiri, N.R. Nair, R.P. John, Bioresour. Technol. **101**(22), 8493–8501 (2010)

Chapter 4
Voided Double-Gyroid Thin Film Templates

The aim of this part of the study was to establish a fabrication route for organic, mesoporous thin film templates of sufficiently large area and homogenous micrometer thickness using the synthesized PXS-*b*-PLA diblock copolymers. The pores of such templates should be connected and continuous from the air/template interface down to the substrate, enabling the replication with functional materials. As discussed previously, for various reasons the most favorable copolymer nanomorphology as starting material for mesoporous templates is the double-gyroid. Hence, is was tested which of the numerous prepared copolymers adopts the double-gyroid morphology.

Preparation of the samples included the following steps, surface modification of the substrate, polymer film deposition, thermal annealing, and selective degradation of the PLA phase, as illustrated in Fig. 4.1. The PLA removal does not alter the copolymers morphology and is comparatively easy, whereas the substrate choice, the surface modification, and thermal annealing protocol decisively influence the copolymer self-assembly behavior as will be discussed in the context of polymer phase separation [2, 3]. The vast possible parameter space of the latter two preparation steps, combined with the numerous copolymers, presented a real challenge.

4.1 Introduction to Block Copolymer Self-Assembly

A remarkable property of polymer melts is their ability to self-assemble, driven by thermodynamic incompatibilities of the different monomers. A brief introduction to the thermodynamic theory of macrophase separation in homopolymer blends and microphase separation in diblock copolymer melts is given. In particular, the effect of controllable parameters, including the monomer interactions, the block composition,

---

Parts of this chapter were previously published [1].

M. R. J. Scherer, *Double-Gyroid-Structured Functional Materials*,
Springer Theses, DOI: 10.1007/978-3-319-00354-2_4,

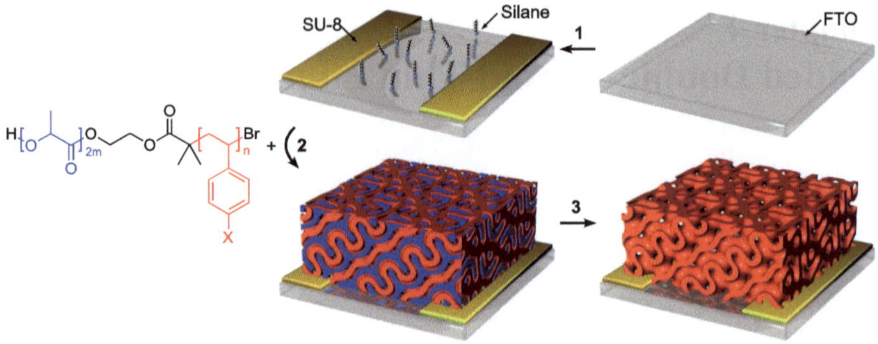

Fig. 4.1 Preparation of mesoporous thin film polymer templates on suitable substrates. **a** Substrate surface modification tailoring a neutral surface. **b** Copolymer film deposition and phase transition to the double-gyroid morphology during thermal annealing. **c** Selective degradation of PLA yields the mesoporous template

the degree of polymerization, the polydispersity and the purity, on the formation of intermaterial dividing surfaces will be discussed.

4.1.1 Macrophase Separation of Homopolymer Blends

Most homopolymer blends tend to segregate or macrophase separate on lowering the temperature T, although mixtures of the corresponding monomers form a single homogeneous phase. The Flory-Huggins mean field theory provides a framework which explains this discrepancy by taking the molecular size of polymers, i.edegree of polymerization N, into account [4].

Considering a binary blend composed of homopolymer A and B with volume fraction f_A and $f_B = 1 - f_A$ and assuming monomers of equal molecular size, the Flory-Huggins lattice model delivers an expression for the Gibbs free energy of mixing

$$\Delta G_{\text{mix}} = \Delta H_{\text{mix}} - T\Delta S_{\text{mix}} \tag{4.1}$$

$$\frac{\Delta G_{\text{mix}}}{k_B T} = \underbrace{\chi f_A f_B}_{\text{enthalpic term}} + \underbrace{f_A N_A^{-1} \ln f_A + f_B N_B^{-1} \ln f_B}_{\text{entropic term}}, \tag{4.2}$$

where k_B is the Boltzmann constant and χ the Flory-Huggins interaction parameter [4]. For $\Delta G_{mix} > 0$ the melt phase separates spontaneously, while it favors a homogenous phase for $\Delta G_{mix} < 0$. Depending on the value of χ, the enthalpic contribution, which is assumed to arise only from nearest-neighbor monomer interactions, can be either positive (opposes mixing) or negative (promotes mixing). In contrast, the entropic term is always negative and thus favors homogeneous mixing.

By scaling the entropy term by a factor of N^{-1} the model accounts for the fact that the monomers are covalently bonded in groups of N, preventing the independent dispersion on lattice sides. Therefore, blend phase behavior is controlled by the product χN, the segregation strength, capturing the two opposing enthalpic and entropic contributions [5].

In addition, the model delivers an expression for the dimensionless interaction parameter χ, which describes the free-energy cost per monomer of contacts between monomer A and B

$$\chi = \frac{Z}{k_B T}\left(\epsilon_{AB} - \frac{1}{2}(\epsilon_{AA} + \epsilon_{BB})\right), \tag{4.3}$$

where Z is the number of nearest-neighbor contacts, and ϵ is the interaction energy per monomer [5]. However, in practice χ may have a more complex dependence on the temperature. Usually it is found to vary linearly with the inverse temperature

$$\chi = a + \frac{b}{T}, \tag{4.4}$$

where the quantity a can be thought of as an entropic component accounting for non-combinatorial entropy changes, while b is the positive or negative enthalpic part expressing an upper or lower critical solution temperatures, respectively. For example, experimentally determined parameters for poly(styrene)/poly(lactic acid) are $\chi = 98.1T^{-1} - 0.112$ and $\chi = 96.6\,T^{-1} - 0.091$ [6].

A phase diagram of the blend showing regions of stability, metastability, and instability, separated by the binodal and spinodal curve, can be derived from the temperature dependence of ΔG_{mix} by applying the common tangent approach. Here only a symmetric polymer blend ($N = N_A = N_B$ and $f = f_A = f_B$) is considered, for a more detailed discussion the reader is referred to the literature [7]. Differentiating Eq. (4.2) twice with respect to f and equating to zero gives a critical value of $\chi N = 2$. Thus, for values $\chi N < 2$ all volume fraction compositions are stable, while for values $\chi N \geq 2$ mixtures of certain compositions will undergo phase separation [7]. For unconnected monomers the entropic terms often overwhelms the enthalpic term and the blend forms a single homogeneous phase. In mixtures of polymers with high N already very weak repulsive monomer interactions lead to a free energy of mixing that favors phase separation.

4.1.2 Microphase Separation of Diblock Copolymers

As in the case of macrophase separation, the two competing terms of enthalpic and entropic nature govern the thermodynamics of block copolymers. Similar to binary polymer bends the phase behavior of diblock copolymer melts is primarily controlled by the segregation strength χN and the volume fractions f_A and $f_B = 1 - f_A$ of the two blocks. Being covalently bonded, repulsive blocks are prevented from separating on a macroscopic level. Hence, the segregation into A- and B-rich domains occurs on the nanoscale, therefor termed microphase separation. This self-assembly process gives rise to the formation of a variety of highly ordered morphologies. Which structure diblock copolymers adopt at equilibrium is highly dependent on the composition, while the dimensions of the microdomain structure depend on the copolymer degree of polymerization N, as schematically shown in Fig. 1.2.

4.1.2.1 Phase Diagrams of Diblock Copolymers

Symmetric diblock copolymers ($f_{\mathrm{A}} \simeq f_{\mathrm{B}}$) arrange into a lamellar morphology, with alternating layers of the constituent blocks. Asymmetric diblock copolymers ($f_{\mathrm{A}} \neq f_{\mathrm{B}}$) self-assemble into morphologies where the minority component is enclosed by the majority or matrix phase. Between the lamellar phase and the hexagonally packed cylinders the bicontinuous $Fddd$ and tricontinuous double-gyroid phase is observed, as previously discussed in Chap. 2. A further increase in asymmetry leads to the formation of body-centered cubic and close-packed spheres.

At sufficiently high temperatures (small χ) the entropic terms exceed the energetic interactions, and the chains are mixed homogeneously. In this case the free energy per chain can be approximated by the A-B contact energy [5]

$$\frac{G_{\mathrm{DIS}}}{k_{\mathrm{B}}T} \approx \chi f_{\mathrm{A}} f_{\mathrm{B}} N. \tag{4.5}$$

As the temperature is reduced (large χ) the repulsive enthalpic term grows and the system tends to minimize the unfavorable interface between the incompatible blocks via formation of intermaterial dividing surfaces (IMDS). For any given segregation strength and copolymer composition the melt adopts the IMDS that minimizes the free energy. During the IMDS formation the polymer chains have to be significantly stretched at the linkage between the blocks. Compared to homopolymer blends, where the chains have random coil conformation, the enthalpic forces now have to counterbalance additional entropic forces arising form the extended chain conformation. Considering a simple model where a chain of N monomers of size a are extended to a distance R, the entropic force law is assumed to be approximately Hookian and hence, can be expressed as $F_{\mathrm{stretch}} = 3k_B T R^2/(2Na^2)$ [5]. Further, the free energy for a symmetric diblock ($f_{\mathrm{A}} = f_{\mathrm{B}}$), which will microphase separate into an alternating lamellar morphology, is given by [5]

$$\frac{G_{\mathrm{LAM}}}{k_{\mathrm{B}}T} = \frac{3(\lambda/2)^2}{2Na^2} + \frac{\gamma A}{k_B T} = \frac{3(\lambda/2)^2}{2Na^2} + \frac{Na\sqrt{\chi/6}}{\lambda/2} \tag{4.6}$$

The first term in Eq. (4.6) is the entropic stretching penalty derived under the assumption that the chains are uniformly stretched to a length of one-half of the lamellar domain period λ. The second term represents the repulsive energetic interactions confined to the (sharp) A-B interface. It is given by the product of the contact area per chain derived under volume filling constraint, $A = Na^3/(\lambda/2)$, and the interfacial tension determined according to classical theory of polymer-polymer interactions, $\gamma = (k_B T/a^2)\sqrt{\chi/6}$. Differentiation of Eq. (4.6) with respect to λ and equating to zero yields an expression for the equilibrium lamellar period [5]

$$\lambda_0 \approx 1.03 \cdot a\chi^{1/6} N^{2/3} \sim N^{2/3}. \tag{4.7}$$

The corresponding free energy of the lamellar phase can be used to estimate a critical value $(\chi N)_{\mathrm{ODT}}$ where the transition from a single homogeneous melt to heterogeneous ordered microdomains occurs, which is known as the order-disorder transition (ODT). For a lamellae-forming, symmetric copolymer ($f_A = f_B$) equating $G_{\mathrm{LAM}}(\lambda_0)$ to $G_{\mathrm{DIS}} \approx k_B T \chi N/4$, compare Eq. (4.5) for the symmetric case, delivers a critical value of $(\chi N)_{\mathrm{ODT}} \approx 10.5$. This value is close to the one derived by more sophisticated self-consistent field theories [8, 9]. For $\chi N > 10.5$, which corresponds to either high molecular weight copolymers or strong incompatibility in combination with low temperatures, symmetric diblock copolymers are predicted to microphase separate into ordered lamellar with period λ_0. Respectively, for $\chi N < 10.5$ a single homogeneous phase, i.ea disordered phase, is expected.

Depending on how distinct the interfaces between the segregated domains are and how pure the composition in the separated domains is, the extent of the block segregation is divided into three regimes: weak ($\chi N \sim 10$), intermediate ($\chi N \sim 10-100$) and strong ($\chi N \geq 100$) segregation limit. For each regime, mean field and analytical theories have been developed. The weak and strong segregation limit theories for block copolymer melts have been unified by Matsen et al. [8, 9]. Their approach involves the numerical solution of self-consistent field equations. By comparing the free energies for different phases, the theoretical phase diagram displayed in Fig. 4.2a is obtained.

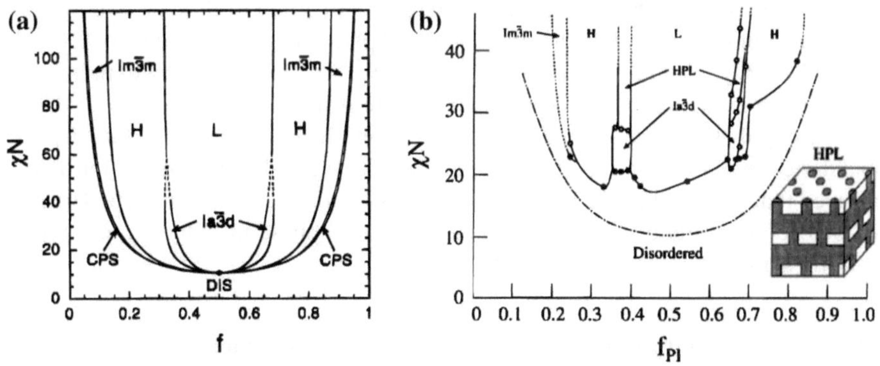

Fig. 4.2 Early diblock copolymer phase diagrams: disordered *DIS*; lamellar *L*; double-gyroid, $Ia\bar{3}d$; hexagonally packed cylinders *H*; body-centered cubic spheres, $Im\bar{3}m$; and close-packed spheres *CPS*. **a** Calculated phase diagram using self-consistent mean field theory (Adapted with permission from Matsen et al. [9]. Copyright 2013 American Chemical Society). **b** Experimentally determined phase diagram for PS-*b*-PI diblock copolymers revealing the existence of the metastable hexagonally perforated lamellar (HPL) phase (Adapted with permission from Khandpur et al. [10]. Copyright 2013 American Chemical Society)

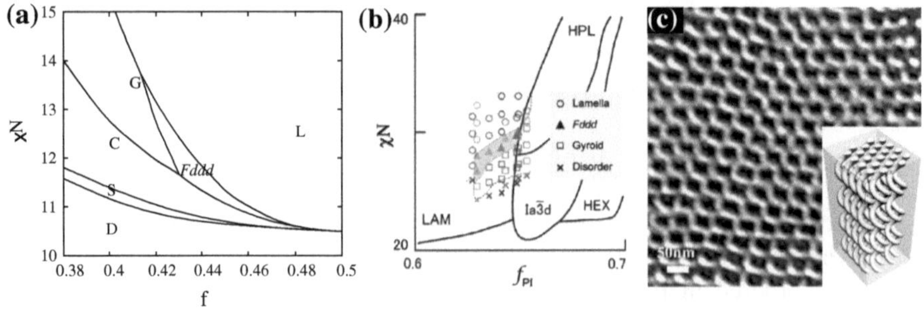

Fig. 4.3 **a** Calculated phase diagram using self-consistent mean field theory predicting the existence of the *Fddd* morphology in diblock copolymers (Reprinted with permission from Tyler et al. [13]. Copyright 2013 by the American Physical Society). **b** Experimentally determined phase diagram for PS-*b*-PI diblock copolymers showing the boundaries of the *Fddd* phase (Reprinted with permission from Kim et al. [14]. Copyright 2013 American Chemical Society). **c** TEM image of helical cylinder phase observed in poly(styrene)-*b*-poly(L-lactic acid) and inset showing a simulation (Reprinted with permission from Ho et al. [15]. Copyright 2013 American Chemical Society)

The theoretically calculated phase diagrams of the mid 1990s agree remarkably well with the experimentally observed phase behavior of diblock copolymers. Figure 4.2b shows the almost symmetric phase diagram of PS-*b*-PI measured by Khandpur et al. in 1995 [10]. At that time it was still believed, that the semi-continuous hexagonally perforated lamellar phase, a combination of the lamellar and cylindrical phase, is a stable equilibrium phase. Later experimental and theoretical studies showed, that the HPL is a metastable structure or long-lived transition, but not an equilibrium structure [11]. Similarly, the *Fddd* was believed to exists as a metastable structure [12].

In 2005, Tyler et al. were the first to predict the existence of the orthorhombic *Fddd* as an equilibrium structure in the phase diagram of diblock copolymers. Using self-consistent field theory, the calculated phase diagram is shown in Fig. 4.3a [13, 16]. Two years later, Takenaka et al. were the first to deliver the experimental proof for the existence of the *Fddd* microdomain structure, discovered in a PS-*b*-PI diblock copolymer melt [17]. Thereafter, Kim et al. determined the *Fddd* phase boundaries in PS-*b*-PI melts, as illustrated in Fig. 4.3b [14, 18]. Although it was believed that the latest phase diagram of diblock copolymer was complete, Ho et al. in 2009 reported the discovery of a helical cylinder phase in poly(styrene)-*b*-poly (L-lactic acid) melts, see Fig. 4.3c [15].

4.1.2.2 Phase Diagrams of Polydisperse Diblock Copolymers

So far, the discussion on copolymer melt behavior was limited to monodisperse blocks, neglecting the fact that most polymers are polydisperse. The polydispersity index is a measure for the width of the molecular weight distribution $PDI = M_w/M_n$. A PDI for the entire copolymer as well as for both blocks PDI_A and PDI_B can be defined. In a recent publication on the effects of polydispersity on

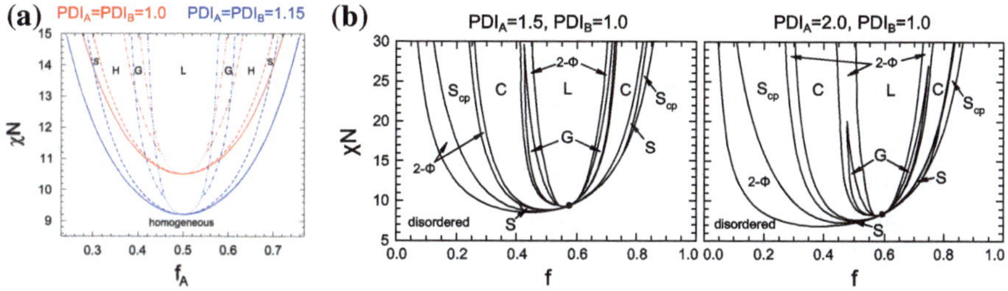

Fig. 4.4 Theoretical diblock copolymer phase diagrams for different levels of polydispersity. **a** An increase of the overall PDI with identical block PDIs results in a shift towards lower χN values while the phase diagram stays symmetrical (Adapted with permission from Cooke et al. [20]. Copyright 2013 American Chemical Society). **b** An asymmetrical increase of PDI_A with $PDI_A > PDI_B$ shifts the phase boundaries towards decreasing χN and increasing f_A values, and creates biphasic regions 2-Φ. Note that the phase space of the pure gyroid morphology is even narrower for asymmetric block polydispersities (Reprinted with permission from Matsen et al. [21]. Copyright 2013 by the American Physical Society)

block copolymer self-assembly Lynd et al. summarize the ongoing research [19]. In Fig. 4.4 the main results from Cooke et al. and Matsen are displayed [20, 21]. Applying self-consistent field theory they predict that changes in the molecular weight distributions of the A and B block alter the entropic and enthalpic free energy balances and thus, the locations of the phase boundaries. In addition, they find the emergence of significant two-phase coexistence regions at the expense of the gyroid phase window. Furthermore, they find evidence that polydispersity relieves packing frustration, which reduces the tendency for long-range order.

Several experimental studies on the effects of the PDI on diblock copolymer self-assembly have appeared in the literature. Bendejaq et al. synthesized several poly(styrene)-*b*-poly(acrylic acid) copolymers with broad molecular weight distributions and found that these systems produced well-ordered structures [22]. Blends of a homopolymer hA and a diblock copolymer A-B can also be considered as a pure copolymer system with large polydispersity. These systems were extensively studied, for more details on this topic the reader is referred to my diploma thesis [23].

Furthermore, Lynd and Hillmyer prepared several sets of poly(ethylene-*alt*-propylene)-*b*-poly(D,L-lactic acid) with controlled molecular weights, compositions, and polydispersity indices to prove the discussed theoretical predictions [24]. In particular, they were interested in the influence of changes in the molecular weight distribution of one or both of the two blocks on the domain sizes and morphological boundaries. They found, that the domain spacing increased with increasing polydispersity. Furthermore, they demonstrated that increasing the polydispersity of the lactide block of otherwise identical polymers can result in a morphology change. More specifically, an increase in the polydispersity of the majority poly(lactic acid) segment of a diblock copolymer gives rise to phase transitions to structures with smaller mean interfacial curvature, see Sect. 2.1. Conversely, when the poly(lactic acid) block is the minority component, a polydispersity increase results in a morphology change possessing a larger mean interfacial curvature. The molecular weight

Table 4.1 Diblock copolymers used in the study by Lynd et al.

Sample ID	N_{total}	f_{PLA}	PDI	PDI $_{PLA}$	M_n^{PLA} (kg/mol)	M_n (kg/mol)	Phase
PL(0.34, 1.18)-4	37.2	0.344	1.08	1.18	1.56	3.56	Lamellae
PL(0.34, 1.37)-4	36.9	0.339	1.15	1.37	1.53	3.53	Gyroid
PL(0.34, 1.43)-4	36.2	0.326	1.17	1.43	1.44	3.44	Cylinder

An increased PDI of the minority block ($f_{PLA} < 0.5$) can cause transitions to morphologies with increased mean interfacial curvature (e.g. lamella → gyroid → cylinder) (Adapted with permission from Lynd et al. [24]. Copyright 2013 American Chemical Society)

properties of three diblock copolymers that show this specific behavior are displayed in Table 4.1. All polymers have similar overall molecular weights and volume ratios, but adopt different microdomain morphologies resulting from the polydispersity increase of the minority poly(lactic acid) block.

4.1.2.3 Temperature-Induced Phase Transitions

Amongst others the overall degree of polymerization, the block composition, the block volume ratio, and the polydispersities determine the phase behavior of diblock copolymers. After polymerization these parameters are fixed and can only be changed indirectly, e.g. by adding homopolymers, preparing mixtures of copolymers, or by solvent annealing with selective solvents [25, 26]. Although it is possible to induce a phase transition by these indirect manipulations they are experimentally challenging and preparation of large area films of good quality is ambitious. However, since the interaction parameter χ is highly temperature dependent it is possible to induce morphology transitions through thermal annealing, which corresponds to an almost vertical path in the phase diagram. Given an inverse temperature dependence, an increase in temperature will result in a shift towards smaller χN values in the phase diagram. Typically, a morphology transition is only achievable when the annealing temperature exceeds the copolymer's glass transition temperature, the experimentally determined values of which vary between 104 °C for PS/PFS and 57 °C for PLA.

The advantage of this technique is that large area thin films of homogenous thickness can be spin or blade coated before the desired morphology is induced by thermal annealing while obtaining a good film quality. Due to fast solvent evaporation during spin coating, the polymer films are trapped in a non-equilibrium state possessing a non-equilibrium phase morphology or a phase morphology close to thermodynamics equilibrium, but without long-range order. To erase any effects of this quenched state on the temperature-induced phase transition, the temperature is typically increased until the copolymers form a disordered phase. Kinetics of morphological transitions in microphase separated diblock copolymers were also studied by computer simulations [12, 27–29].

However, inducing phase transitions to a desired phase by thermal annealing also faces some limitations. Most polymers only show a limited temperature stability and start to degrade or depolymerize at elevated temperatures [30–32]. Especially, for

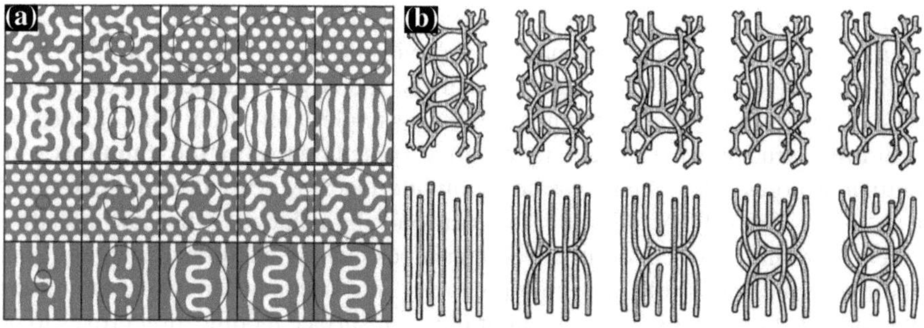

Fig. 4.5 Order-order transition from cylinders to double-gyroid and vice versa. **a** The minimum energy path connecting the two ordered phases is computed using the recently developed string method within the self-consistent field theory (Reprinted with permission from Cheng et al. [29]. Copyright 2013 by the American Physical Society) **b** Schematic illustration of nucleation and growth (Reprinted with permission from Matsen et al. [27]. Copyright 2013 by the American Physical Society)

high molecular weight polymers phase transition occurring at low χN values are not feasible since they require very high annealing temperatures. Another limitation is the extended annealing time required for high molecular weight copolymers to reach equilibrium. For homopolymer melts the de Gennes' self-diffusion model predicts a reptation time of $\tau \approx N^3$ [7]. Assuming that this relation also holds for diblock copolymer it can be rewritten using Eq. (4.7) as $\tau \approx \lambda_0^{4.5}$. When attempting to double the unit cell dimension a morphological transition will require a more than 20-fold increase in annealing time. Furthermore, some studies indicate that due to localization caused by the presence of interfaces in microphase separated copolymer melts the reptating chains are significantly retarded, which would require even longer annealing times [33] (Fig. 4.5).

4.1.2.4 Microphase Separation in Thin Films

So far, only bulk behavior was considered, neglecting any energetic contributions arising from interfaces. Especially, in thin films the interactions between the constituent blocks with both the free surface and the substrate generally lead to a different phase behavior. When the substrate's surface chemistry favors the preferential wetting by one of the blocks, cylinders or lamellae orientated parallel to the substrate are typically formed. This can either lead to the formation of a wetting layer not thicker than the end-to-end length of the copolymer or even affect the morphology adopted in the bulk of the thin film. Relatively few studies investigated the thin films behavior of block copolymers that adopt gyroid-like morphologies in the bulk [34]. Two theoretical studies employing computer simulations suggest, that thin film effects frequently, although not always, destabilize the gyroid morphology [35, 36]. Thin films of PS-*b*-PI diblock copolymers, and PS-*b*-PI/hPS blends solvent cast onto silicon adopted the gyroid morphology [11, 37–39], but when cast onto transparent conductive fluorine-doped tin oxide, the PS-*b*-PI systems did not form wetting PS layer [23, 40]. Epps et al. examined the behavior of PI-PS-PEO in thin films with thicknesses of less than twice the bulk domain periodicity. While the triblock copolymers adopt a gyroid-like morphology in bulk samples, very thin films formed no gyroid-like structures [41]. The only reported voided double-gyroid thin films with continuous pores extending across both film interfaces, were reported by Crossland et al. using PFS-*b*-PLA [42–46]. This can only be achieved when the substrate presents a neutral surface for all copolymer blocks taking additional energetic contributions from a possible surface topography into account. However, tuning the substrates surface energy by chemical or physical surface modification proves to be challenging.

4.1.2.5 Concluding Remarks

A number of parameters determine whether or not the morphology of interest is adopted by a microphase separating copolymer melt. Of these, most important are the interaction parameter, the volume ratio of the blocks, the degree of polymerization, the individual block molecular weight distributions, the overall polydispersity, the interactions with the interfaces, and last but not least, the temperature. Only the latter two parameters are experimentally accessible and can be altered after the synthesis of the copolymer (disregarding polymer blends). Control over the self-assembly at the film interfaces becomes essential when the polymer films are intended to be used as templates. Meuler et al. recently published a comprehensive review on how these various parameters affect the formation of gyroid-like morphologies in polymeric materials [47].

4.2 Experimental Methods

4.2.1 Materials

Besides the synthesized copolymers, the chemicals and materials listed in Table 4.2 were used as received unless stated otherwise.

4.2.2 Substrate Surface Modification

The here presented surface treatment procedure was the outcome of a tedious optimization process that will be discussed in Sect. 4.3.6. FTO-coated glass substrates were cleaned for 15 min at 80 °C in a strong piranha solution, containing three parts laboratory reagent grade sulfuric acid and one part hydrogen peroxide. The substrates were heated in the sulfuric acid before adding the hydrogen peroxide. Thereafter, the substrates were rinsed thoroughly and sonicated for 3 min using DI water, and dried

Table 4.2 Chemicals and materials used for the template preparation

Chemical	Purity (%)[a]	Abbreviation
Bis(trimethylsilyl)amine	97 (SA)	HMDS
Chloroplatinic acid	99.9 (SA)	–
Cyclohexane, anhydrous	99.5 (SA)	–
Fluorine-doped tin oxide coated glass, 15 Ω/□	– (S)	FTO
Hydrogen peroxide	30 (SA)	–
Methanol, laboratory reagent grade	– (FS)	–
Methyltrichlorosilane	97 (SA)	MTS
Octyltrichlorosilane	97 (SA)	OTS
Propylene glycol monomethyl ether acetate	99 (SA)	PGMEA
Sodium hydroxide	– (SA)	–
SU-8 2000.5	– (MC)	–
Sulfuric acid, laboratory reagent grade	95 (FS)	–
Sulfuric acid	99.999 (SA)	–
Toluene, anhydrous	99.8 (SA)	–
Water, deionized 18 MΩ	–	DI water

[a] Suppliers are given in brackets: Fisher Scientific *FS*; MicroChem *MC*; Sigma Aldrich *SA*; Solaronix *S*

on a hotplate at 100 °C. A fresh 4.3 mM octyltrichlorosilane solution was prepared in a flame dried glass beaker by adding OTS to anhydrous cyclohexane. Directly before immersion into the silanization solution the substrates were quenched to room temperature by placing them on a cool metal plate using dry metal tweezers [48]. After around 10 s the substrates were taken from the solution and immediately dried under a strong nitrogen stream. This treatment gave rise to a water contact angle of $59 \pm 5°$. The time between piranha cleaning and silanization was kept as short as possible to prevent reabsorption of airborne hydrocarbons.

The cleaned and silanized substrates were selectively patterned by photolithography using an electrically insulating photoresist [49]. A thin SU-8 2000.5 photoresist layer was spun at 6,000 rpm, soft-baked, exposed using a MJB4 mask aligner (60–80 mJ cm^{-2}), post-baked, and developed in PGMEA for 30 s [49]. The applied soft and post-bake procedure was as follows: 1 min at 65 °C, 1 min at 92 °C, and 1 min at 65 °C.

4.2.3 Thin Film Deposition

The synthesized diblock copolymers were dried in a vacuum oven at 60 °C overnight. Then 10 $\%_{wt}$ stock solutions were prepared in anhydrous toluene and filtered with 0.45 μm Teflon (PTFE) filters. Micrometer thick copolymer films were spun, drop- or blade-coated, typically using 3.5 μl of copolymer solution per square centimeter. Films prepared by spin coating were spun at 200–700 rpm for 45 s.

Fig. 4.6 Homemade oven with lid set up inside a conventional vacuum oven for precise control over the heating rate and annealing temperature

4.2.4 Thermal Annealing

The optimization process that led to the here presented DG-yielding thermal annealing procedure will be discussed in Sect. 4.3.4. To enable precise control over the temperature with fast heating rates for the thermal annealing process a 'mini-oven' was designed for use inside a conventional vacuum oven. With the lid closed the mini-oven guarantees a homogeneous temperature throughout the whole film under an inert gas atmosphere as well as under vacuum.

Both ovens were preheated to 90 °C. The conventional oven was kept at this temperature, while the mini-oven was used to control the annealing temperature. After placing the samples inside the oven the annealing chamber was alternatingly evacuated and refilled with nitrogen for several times in order to prevent polymer degradation through oxidation. The optimized 'standard annealing' protocol was as follows: The mini-oven was heated up to 150 °C at a ramp rate of 2.5 °C min^{-1} under vacuum giving the polymer films sufficient time to dry. On reaching 150 °C the oven was refilled with nitrogen and kept under a constant flow of nitrogen for the remaining annealing time. To prepare double-gyroid-structured films for the subsequent usage as templates, an annealing time of 20 min at 173 °C or 10 min at 205 °C depending on the copolymer's molecular weight was chosen. Finally, the samples were quenched to room temperature by placing them on a cool metal plate (Fig. 4.6).

4.2.5 Selective PLA Degradation

For the selective removal of the poly(lactic acid) phase the samples were immersed in a 0.1 M aqueous sodium hydroxide solution containing 50 $\%_{vol}$ methanol for several hours [6, 50]. The resulting mesoporous polystyrene templates were soaked and washed in methanol for several minutes to remove the base and then carefully dried with a nitrogen stream.

4.2.6 Scanning Electron Microscopy

Scanning electron micrographs (SEMs) of the voided polymer templates were taken on a LEO ULTRA 55 SEM with a Schottky-emitter (ZrO/W cathode) and an in-lens detector. The acceleration voltage was typically set in a range of 3–10 keV. In order to prevent the electrically insulating poly(styrene) samples from charging up, they were mounted onto SEM stubs with conduction carbon tape and were coated with a few nanometer thick platinum or gold layer using a sputter coater. Cross-sections were prepared by simply breaking the films after PLA degradation.

4.3 Results and Discussion

The substrate surface treatment and thermal annealing protocol used to obtain double-gyroid-structured MHI41 films with a porous film/substrate interface were found to be unsuitable for the diblock copolymers synthesized in this study [44]. An inappropriate surface treatment and thermal annealing can individually prevent the double-gyroid from being the thermodynamically stable morphology. This complicated the search for the right polymerization conditions and made the development of the here presented synthesis and fabrication protocol for the double-gyroid templates a tedious iterative process.

To make matters even worse, the vacuum ovens used before installing the mini-oven were falsely calibrated and had a temperature offset of 10 and 15 °C. Combined, these circumstances made initial experiments poorly reproducible. Nonetheless, almost five dozen of the synthesized diblock copolymers were identified to form double-gyroids. They are listed in Table 4.3. The complete list of all studied copolymers together with their theoretical PLA volume fraction f_{PLA}^{theo} and the adopted micromorphology for selected thermal annealing conditions are given in Table B.1. In the following a few, elective experiments that led to the determination of the appropriate experimental conditions, chosen from a large number of unsuccessful experiments, are presented. The SEM images presented here are reprinted in larger size in Appendix C.

4.3.1 Degradation During Thermal Annealing

Poly(styrene) is fairly temperature stable in contrast to PLA which starts to thermally degrade at 180 °C even under an inert atmosphere [31, 32, 51]. The thermal

Table 4.3 Microphase separated structure adopted by PXS-*b*-PLAs for different thermal annealing conditions

Sample	M_n^{PXS} $\left[\frac{\mathrm{kg}}{\mathrm{mol}}\right]^{\mathrm{a}}$	m_{PXS} [mg][b]	$f_{\mathrm{PLA}}^{\mathrm{theo}}$ [%][c]	$\mathrm{Structure}_t^T$ $[^{\circ\mathrm{C}}_{\mathrm{h}}]^{\mathrm{d}}$
PFS18-*b*-PLA32 [e]	15	25	42.50	$\mathbf{DG}_{19}^{170}$, W_{-}^{-}
PFS18-*b*-PLA33 [e]	15	25	45.00	${}^{\star}\mathbf{DG}_{19}^{170}$
PFS18-*b*-PLA47 [e]	15	50	41.00	X_{19}^{170}, $[\mathbf{DG},\mathrm{C}]_{20}^{190}$, $\mathbf{DG}_{20}^{170}$
PFS18-*b*-PLA48	15	50	43.00	$[\mathrm{X},{}^{\parallel}\mathrm{C}]_{19}^{170}$, $[{}^{\star}\mathbf{DG},\mathrm{HPL},{}^{\perp}\mathrm{C}]_{20}^{165}$
PFS18-*b*-PLA75	15	50	41.00	${}^{\perp}\mathrm{C}_{20}^{190}$, $[\mathbf{DG},{}^{\perp}\mathrm{C}]_{17}^{170}$
PFS18-*b*-PLA76 [ef]	15	50	43.50	$[{}^{\star}\mathbf{DG},\mathrm{HPL},{}^{\perp}\mathrm{C},\mathrm{X}]_{17}^{165}$, ${}^{\perp}\mathrm{C}_{20}^{190}$, HPL_{39}^{100}, $\mathrm{C}_{0.25}^{200}$, X_{17}^{170}
PFS18-*b*-PLA81	15	50	41.00	HPL_{-}^{-}, $[{}^{\star}\mathbf{DG},\mathrm{HPL},{}^{\perp}\mathrm{C}]_{20}^{190}$, ${}^{\star}\mathbf{DG}_{2}^{170}$
PFS18-*b*-PLA82 [f]	15	50	45.00	HPL_{48}^{150}, $[{}^{\star}\mathbf{DG},\mathrm{HPL},{}^{\perp}\mathrm{C},\mathrm{X}]_{20}^{190}$
PFS18-*b*-PLA96	15	35	40.00	$\mathbf{DG}_{39}^{160}$
PFS18-*b*-PLA97 [f]	15	35	41.25	$\mathbf{DG}_{39}^{160}$
PFS18-*b*-PLA98	15	35	42.50	$\mathbf{DG}_{39}^{160}$
PFS26-*b*-PLA118 [f]	17	500	38.00	W_{24}^{120}, ${}^{\perp}\mathrm{C}_{24}^{190}$, $\mathbf{DG}_{1}^{170}$, $\mathbf{DG}_{1}^{160}$
PFS26-*b*-PLA119 [f]	17	500	39.00	C_{24}^{120}, ${}^{\perp}\mathrm{C}_{24}^{190}$, $\mathbf{DG}_{0.66}^{180}$
PFS26-*b*-PLA120	17	500	40.00	C_{24}^{190}, HPL_{1}^{160}, $[\mathbf{DG},\mathrm{C}]_{0.25}^{210}$
PFS26-*b*-PLA127 [ef]	17	500	38.00	$\mathbf{DG}_{0.66}^{180}$
PFS26-*b*-PLA132 [f]	17	500	38.25	$\mathbf{DG}_{0.33}^{190}$, $\mathrm{W}_{0.75}^{170}$
PFS26-*b*-PLA133 [f]	17	500	38.50	$\mathbf{DG}_{0.33}^{190}$, $\mathrm{W}_{0.75}^{170}$
PFS26-*b*-PLA134	17	750	38.00	$\mathbf{DG}_{0.75}^{180}$, $\mathrm{W}_{0.75}^{170}$, $\mathbf{DG}_{0.25}^{200}$
PFS26-*b*-PLA135	17	750	38.50	$\mathbf{DG}_{0.75}^{180}$, $[\mathrm{W},\mathbf{DG}]_{0.75}^{170}$
PFS26-*b*-PLA136 [e]	17	750	39.00	$\mathbf{DG}_{0.75}^{180}$, $\mathrm{W}_{0.75}^{170}$, $\mathbf{DG}_{0.33}^{190}$, $\mathbf{DG}_{0.25}^{200}$
PFS18-*b*-PLA144	15	500	38.50	$\mathbf{DG}_{0.33}^{190}$, $\mathbf{DG}_{0.33}^{183}$
P(F)S39-*b*-PLA150 [f]	19	500	38.50	$\mathrm{W}_{0.33}^{183}$, $\mathrm{W}_{0.16}^{195}$, $\mathbf{DG}_{0.1}^{210}$
P(F)S40-*b*-PLA151	17	500	38.50	$\mathrm{W}_{0.33}^{183}$, $\mathbf{DG}_{0.16}^{195}$, $\mathbf{DG}_{0.1}^{210}$, $\mathbf{DG}_{0.16}^{195}$
PFS26-*b*-PLA118	17	500	38.00	$\mathbf{DG}_{0.33}^{183}$, $\mathbf{DG}_{0.25}^{175}$
P(F)S39-*b*-PLA155	19	500	38.00	$\mathrm{W}_{0.16}^{181}$, $\mathbf{DG}_{0.1}^{210}$
P(F)S40-*b*-PLA156	17	500	38.00	$\mathbf{DG}_{0.16}^{181}$, $\mathbf{DG}_{0.1}^{210}$
P(F)S41-*b*-PLA157	18	500	38.00	$\mathrm{W}_{0.16}^{181}$, $\mathbf{DG}_{0.1}^{210}$
P(F)S39-*b*-PLA160	19	500	37.50	$\mathbf{DG}_{0.16}^{195}$, $\mathbf{DG}_{0.1}^{210}$
P(F)S40-*b*-PLA161	17	500	37.50	$\mathbf{DG}_{0.1}^{210}$
P(F)S41-*b*-PLA162	18	500	37.50	$\mathbf{DG}_{0.1}^{210}$
PFS18-*b*-PLA163	15	500	38.30	$\mathbf{DG}_{0.33}^{181}$, ${}^{\parallel}\mathrm{C}_{0.16}^{205}$
P(F)S39-*b*-PLA165	19	1000	38.00	$\mathrm{W}_{0.33}^{181}$, $[\mathbf{DG},\mathrm{C}]_{0.16}^{205}$, $\mathrm{C}_{0.05}^{235}$
P(F)S40-*b*-PLA166	17	1000	38.00	$\mathrm{W}_{0.33}^{181}$, $\mathbf{DG}_{0.16}^{205}$, $\mathrm{X}_{0.05}^{235}$
P(F)S41-*b*-PLA167	18	1000	38.00	$\mathrm{W}_{0.33}^{181}$, $\mathbf{DG}_{0.16}^{205}$, $\mathrm{X}_{0.05}^{235}$, $\mathrm{X}_{0.55}^{205}$
P(F)S42-*b*-PLA168	17	500	38.00	$\mathrm{W}_{0.33}^{181}$, $\mathbf{DG}_{0.16}^{205}$, $\mathrm{X}_{0.05}^{235}$, $\mathrm{X}_{0.55}^{205}$
P(F)S42-*b*-PLA169	17	500	37.50	$[\mathbf{DG},\mathrm{W}]_{0.33}^{181}$, $\mathbf{DG}_{0.16}^{205}$, $\mathrm{X}_{0.05}^{235}$, $\mathrm{X}_{0.55}^{205}$
PFS44-*b*-PLA172	15	500	38.00	$\mathbf{DG}_{0.25}^{180}$, ${}^{\parallel}\mathrm{C}_{0.16}^{205}$, $\mathrm{W}_{0.25}^{170}$
PFS44-*b*-PLA173	15	500	38.50	$\mathbf{DG}_{0.25}^{180}$, $[\mathbf{DG},\mathrm{C}]_{0.16}^{205}$, $\mathrm{W}_{0.25}^{170}$
PFS45-*b*-PLA174	15	500	38.00	$\mathbf{DG}_{0.25}^{180}$, ${}^{\perp}\mathrm{C}_{0.16}^{205}$, $\mathbf{DG}_{0.25}^{170}$

(continued)

Table 4.3 (continued)

Sample	M_n^{PXS} $\left[\frac{kg}{mol}\right]^a$	m_{PXS} [mg][b]	f_{PLA}^{theo} [%][c]	Structure_t^T $[^{\circ C}_{h}]^d$
PFS45-*b*-PLA175	15	500	38.50	$\mathbf{DG}^{180}_{0.25}$, $^{\perp}C^{205}_{0.16}$
P(F)S46-*b*-PLA177	18	500	38.00	$W^{180}_{0.25}$, $\mathbf{DG}^{205}_{0.16}$
PFS44-*b*-PLA183	15	500	39.00	$[HPL,\mathbf{DG}]^{180}_{0.25}$, $[\mathbf{DG},C]^{205}_{0.16}$
PFS45-*b*-PLA184	15	1500	38.25	$\mathbf{DG}^{180}_{0.16}$, $\mathbf{DG}^{180}_{0.25}$, $^{\perp}C^{205}_{0.16}$
P(F)S46-*b*-PLA185	18	500	38.50	$W^{180}_{0.25}$, $[W,\mathbf{DG}]^{205}_{0.16}$
P(F)S49-*b*-PLA192 [e]	14	500	38.25	$\mathbf{DG}^{180}_{0.25}$, $^{\parallel}C^{205}_{0.16}$, $\mathbf{DG}^{175}_{0.25}$
PS50-*b*-PLA201	13	500	37.25	$\mathbf{DG}^{180}_{0.25}$, $^{\perp}C^{205}_{0.16}$
PS50-*b*-PLA202	13	500	39.25	$\mathbf{DG}^{180}_{0.25}$, $^{\perp}C^{205}_{0.16}$
PS50-*b*-PLA203	13	500	40.25	$[\mathbf{DG},HPL]^{180}_{0.25}$, $^{\perp}C^{205}_{0.16}$
P(F)S49-*b*-PLA204 [ef]	14	500	38.25	$\mathbf{DG}^{180}_{0.25}$
P(F)S49-*b*-PLA205	14	500	38.25	$\mathbf{DG}^{180}_{0.25}$
PS50-*b*-PLA206	13	500	37.60	$\mathbf{DG}^{180}_{0.25}$
P(F)S41-*b*-PLA207	18	500	39.30	$W^{174}_{0.33}$, $\mathbf{DG}^{210}_{0.1}$
P(F)S49-*b*-PLA209	14	1000	38.25	$\mathbf{DG}^{174}_{0.33}$
P(F)S49-*b*-PLA210	14	1000	38.25	$\mathbf{DG}^{174}_{0.33}$
P(F)S49-*b*-PLA211	14	1000	38.25	$\mathbf{DG}^{174}_{0.33}$
P(F)S49-*b*-PLA212	14	1000	38.25	$\mathbf{DG}^{174}_{0.33}$
PS50-*b*-PLA214 [e]	13	1000	37.50	$\mathbf{DG}^{174}_{0.33}$
PS50-*b*-PLA215	13	1000	38.00	$\mathbf{DG}^{174}_{0.33}$
PFS26-*b*-PLA120:126 = 1:1			38.50	$C^{180}_{0.75}$, $\mathbf{DG}^{170}_{0.25}$
PFS26-*b*-PLA120:PFS29 = 19:1			38.50	$C^{180}_{0.75}$, $\mathbf{DG}^{170}_{0.25}$

[a] Molecular weight of PXS-OH determined by SEC calibrated with poly(styrene) standards.
[b] Amount of macroinitiator PXS-OH used.
[c] Theoretical volume fraction of PLA assuming $\rho_{PLA} = 1.18\,\text{g cm}^{-3}$.
[d] Morphologies observed in microphase separated films after thermal annealing: No structure *X*; spheres *S*; cylinders *C*; cylinders orientated parallel to the substrate $^{\parallel}C$; cylinders orientated perpendicular to the substrate $^{\perp}C$; bicontinuous worm-like structure *W*; lamellae *L*; hexagonally perforated lamellae *HPL*; double-gyroid *DG*; double-gyroid crystals $^{\star}DG$. Coexisting morphologies are given in square brackets. The annealing temperature [°C] is given as superscript and the annealing time [h] as subscript.
[e] For SEC data see Table 3.6.
[f] For NMR data see Table 3.6.

stability of PLA greatly depends on its purity and end-group. Hydroxyl end-groups as well as impurities, such as metal catalysts, lactide monomer, protic solvents, and oxygen, were found to accelerate the thermal degradation [30, 31]. Issues with the thermal stability of PLA also emerged in this study especially during long annealing of the copolymers, see Fig. 4.7. However, the thermal stability across the synthesized copolymers varied significantly, while some polymers were stable at elevated temperatures for several days, others seemed to degrade exceptionally readily such as PFS26-*b*-PLA116.

The observed inconsistencies are most likely do to varying levels of impurities such as copper salts or monomers. Curiously, degraded films produced bubbles during immersion in the etching solution. The protection of the hydroxyl end-group

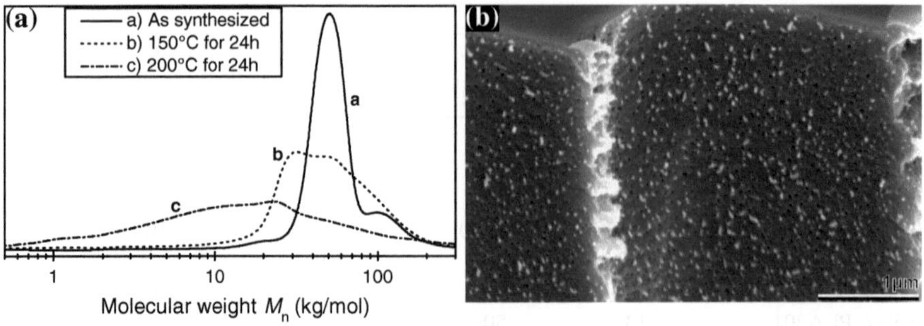

Fig. 4.7 Thermal decomposition of copolymers containing PLA. **a** SEC traces of PFS26-*b*-PLA116 tempered at different temperatures under an inert atmosphere. The PMMA-equivalent molecular weight data was obtained: *a* $M_n = 48.5$, PDI = 1.07; *b* $M_n = 36.4$, PDI = 1.31; *c* $M_n = 12.7$, PDI = 1.95. **b** Cross-sectional SEM of a degraded PFS18-*b*-PLA75$|_{16\,h}^{180\,°C}$ film

via conversion into an acetyl group using acetic anhydride 1.5 $\%_{vol}$ in chloroform was reported to significantly improve the thermal stability [31, 52]. A few attempts were undertaken to acetylate the copolymers shortly before discovering that the DG is adopted by copolymer with a reasonably low molecular weight during brief annealing durations, thereby avoiding thermal degradation and making the acetylation unnecessary. Nevertheless, this protection and a possible purification by liquid-liquid extraction of the copolymer solution in chloroform or DCM with a HCl aqueous solution should be reconsidered for future attempts to obtain 'larger' DG by prolonged high temperature annealing of high molecular weight polymers.

4.3.2 High Molecular Weight Polymers

Although the theoretical phase diagrams suggest the existence of the double-gyroid phase at high χN values, in this study no such structure was observed for copolymers consisting of a styrenic block with a molecular weight M_n larger than 19 kg mol^{-1}. As discussed previously, high molecular weight copolymers require elevated temperatures to increase their mobility and prolonged annealing times to reach the thermodynamic equilibrium morphology. Despite meeting these experimental conditions only spheres, lamellae, hexagonally perforated lamellae, cylinders, and worm-like structures were observed, see Table B.1 and Fig. 4.8. Some of these structures are possibly kinetically trapped and do not constitute the equilibrium morphologies. However, the worm-like structures are also bicontinuous and are therefore equally interesting as template for functional materials. They are subject to ongoing experiments.

Due to the absence of any phase transitions or rearrangement resulting in long-range order, a further increase in annealing time is expected to be less effective, while

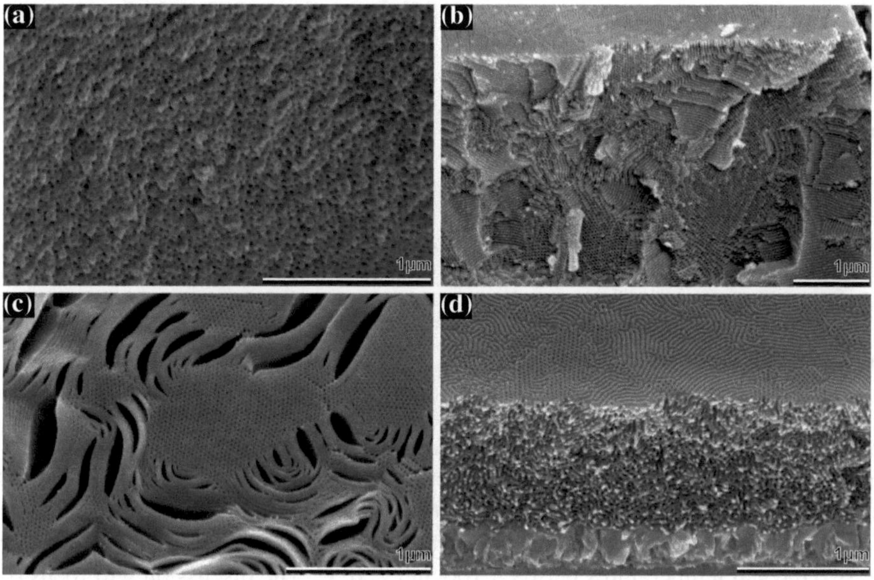

Fig. 4.8 Microphase separated morphologies adopted by high molecular weight copolymers: **a** spheres (PFS19-*b*-PLA35$|_{19\,h}^{170\,°C}$); **b** cylinders with a center-to-center distance of 32 nm (PFS23-*b*-PLA92$|_{20\,h}^{190\,°C}$); **c** coexisting cylinders and lamellae (PFS19-*b*-PLA61$|_{14\,h}^{170\,°C}$); **d**, worm-like structure (PFS27-*b*-PLA128$|_{0.75\,h}^{180°C}$)

prolonged annealing at even higher temperatures proved to result in polymer degradation. Although numerous copolymers with different compositions were synthesized and annealed under varying conditions, it cannot be ruled out that double-gyroids with larger unit cell dimensions can be obtained from high molecular weight PXS-*b*-PLA copolymers by thermal annealing. Other annealing techniques, such as slow solution casting, might have to be employed as demonstrated by Hsueh et al. using PS-*b*-PLA with an overall molecular weight of $M_n = 61\,kg\,mol^{-1}$ [53]. Thus, in the following discussion these high molecular weight copolymers will be neglected

4.3.3 Initial Prolonged Thermal Annealing Attempts

In the following only copolymers containing a PXS block with a molecular weight lower than $20\,kg\,mol^{-1}$ are considered. Initially long annealing times of up to 70 h were tested, as will be reported next. To avoid thermal decomposition and make the template preparation more economical a protocol for rapid annealing with a reduced total annealing time of under one hour was established.

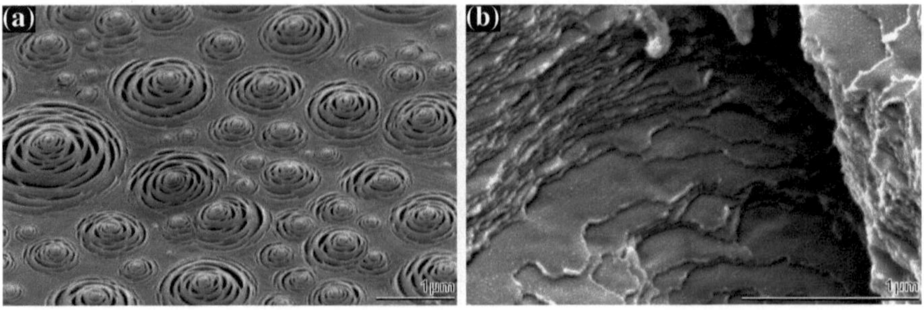

Fig. 4.9 **a** HPL orientated perpendicularly to the substrate (PFS18-*b*-PLA76$|^{150\,°C}_{48\,h}$). **b** HPL aligned parallel to the substrate (PFS18-*b*-PLA48$|^{150\,°C}_{48\,h}$)

The DG-yielding annealing protocol from Crossland et al. optimized for MHI41, namely 180 °C for 35–70 h under vacuum or a nitrogen atmosphere,[1] served as starting point for this study. Initially, the copolymers were spun onto acetone cleaned FTO substrates and dried for at least 3 h at 100 °C under vacuum to remove remaining solvent, water, and methanol. Thereafter, the films were annealed from two hours up to two days at temperatures between 100 and 220 °C. The discussion will focus on polymers that formed double-gyroids with large periodic domains, namely PFS18-*b*-PLA33,48,76,81,82.

4.3.3.1 Prolonged Annealing at 100–150 °C

For annealing temperatures from 100 to 150 °C and annealing times up to 72 h the as-spun morphology was persistent and no phase transition was observed. In films of PFS18-*b*-PLA76 and PFS18-*b*-PLA82 hexagonally perforated lamellae aligned perpendicularly to the substrate ($^{\perp}$HPL) were observed, forming layered cylinders, see Fig. 4.9a. Whereas, ($^{\parallel}$HPL) orientated parallel to the substrate were found in films of PFS18-*b*-PLA48, see Fig. 4.9b.

4.3.3.2 Prolonged Annealing at ≈165 °C

At an annealing temperature of ≈165° locally separated crystallites with double-gyroid morphology evolved via nucleation and growth in HPL films. The nucleation seemed to be preferentially initiated by defects in the FTO or impurities in the film. Round-shaped, fairly single-domain double-gyroid islands emerged in films with $^{\perp}$HPL morphology, see Fig. 4.10. In contrast, films with $^{\parallel}$HPL induced the growth of double-gyroid crystallites forming straight interfaces, see Fig. 4.11. Again the double-gyroid in theses ares were found to be fairly single-crystalline. Another interesting fact is, that the free-surface of the double-gyroid structured film is very smooth compared to the HPL regions.

[1] An annealing temperature of 165–170 °C was rather assumed, since the oven used by Crossland et al. was found to have a faulty calibration.

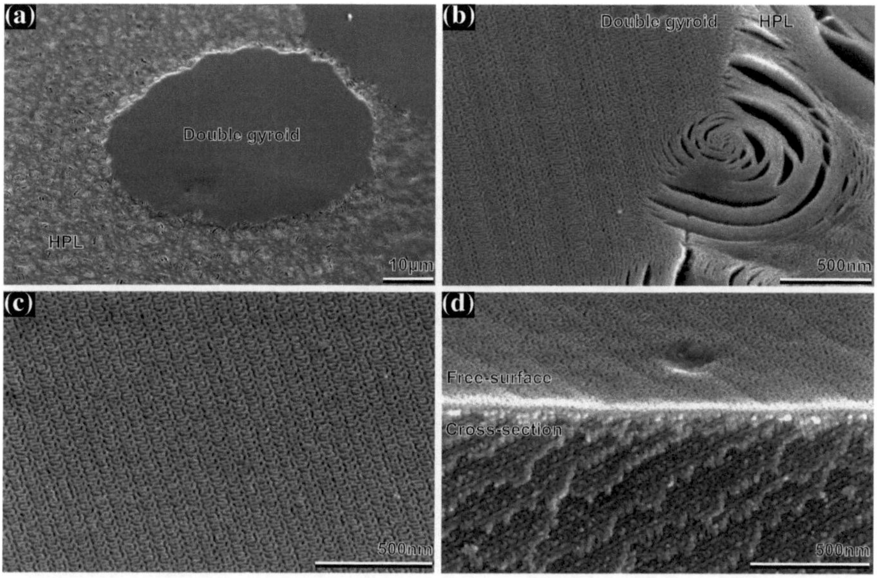

Fig. 4.10 Transition from $^{\perp}$HPL to DG in films of PFS18-*b*-PLA76$|_{17\,h}^{160-165\,°C}$. **a** Top view onto *round shaped*, single-crystalline DG region surrounded by $^{\perp}$HPL. **b** Closeup of (**a**) showing the interface between the $^{\perp}$HPL and DG area. **c** Another magnification of (**a**) showing the free-surface of the fairly single-crystalline DG extending across the whole *round shaped* area. **d** Free-surface and cross-sectional view on single-crystalline DG area

4.3.3.3 Prolonged Annealing at ≈170–200 °C

At annealing temperatures between 170 and 200 °C the morphology transition from HPL to double-gyroid was not limited to a few nucleation sites and therefore large parts of the film were transformed to the double-gyroid morphology. However, before a complete transformation from HPL to gyroid was reached a second phase transition occurred, namely double-gyroid to cylinders orientated perpendicular to the substrate ($^{\perp}$C). Polymer degradation became an issue and the phase transition the cylinders was followed by a transformation into a structureless film. In Fig. 4.12a an elongated defect in the substrate induced a transition from HPL to double-gyroid to cylinders and finally, led to degradation. The three frontiers defined by the interfaces between the different morphologies progressed with increasing annealing time.

Brief annealing experiments with double-gyroid-forming copolymers, as will be discussed in Sect. 4.3.4, suggest that a phase transition from the double-gyroid to a cylindrical morphology can be induced by an increase in annealing temperature. I believe however that the morphology change observed during brief annealing is a real thermodynamic phase transition, whereas the transformation during long annealing at high temperatures might be induced by weight loss of the PLA block through degradation resulting in a more asymmetric copolymer favoring the cylindrical phase.

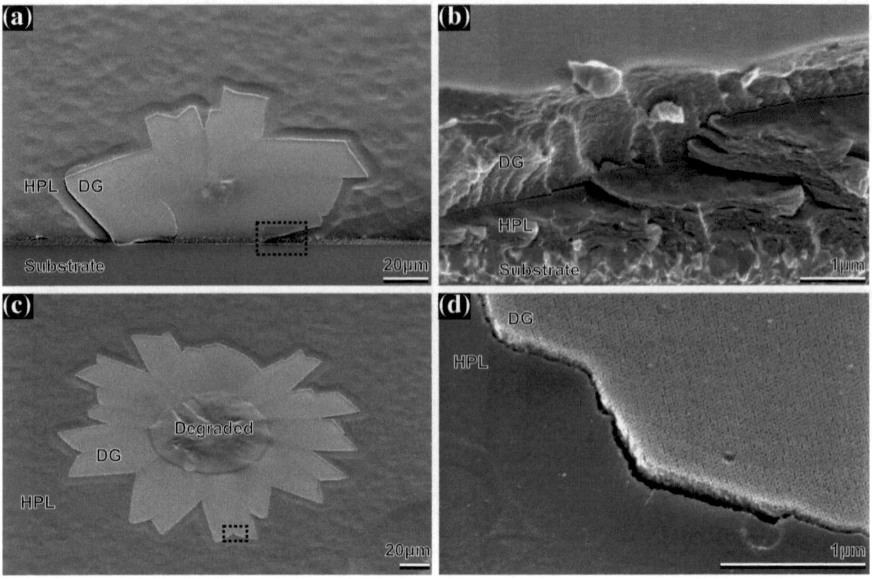

Fig. 4.11 Transition from $^{\parallel}$HPL to double-gyroid in films of PFS18-*b*-PLA48$|_{20\,\mathrm{h}}^{160-165\,^{\circ}\mathrm{C}}$. **a** Top view onto double-gyroid crystals surrounded by $^{\parallel}$HPL which were nucleated by an impurity located in the center. **b** Closeup of the framed area of (**a**) showing a cross-section of the interface between the double-gyroid and $^{\parallel}$HPL. **c** Top view onto a multiple double-gyroid crystals growing from a nucleation point forming a sunflower. **d** Magnified view of the framed area of (**c**)

This hypothesis is supported by the fact that the films seemed to adopted the cylindrical phase faster when annealed under vacuum, driven by an enhanced evaporation of the volatile PLA decomposition products, compared to a nitrogen atmosphere. Further, the PLA volume fraction of PFS18-*b*-PLA33,48,76,81,82 seemed to be at least three percentage points above the one of the copolymers forming the double-gyroid structure during brief annealing at 174 °C, see Sect. 4.3.4 and Table 3.6 and B.1. However, copolymers also containing PFS18, but less PLA, such as PFS18-*b*-PLA46 and PFS18-*b*-PLA30, only exhibited spheres or no structure after the same prolonged annealing procedure, although these had a PLA volume fraction for which double-gyroids formation was expected.

Examination of the cross-section of the film displayed in Fig. 4.12a revealed, that the transition from double-gyroid to cylinders originated at the film's free surface (Fig. 4.12b). From SEM images showing the free-surface of standing cylinders in films of PFS18-*b*-PLA76 a center-to-center distance of approximately 23 nm was determined, which is similar to the value measured for MHI41. With increasing annealing time the cylinders showed signs of degradation before the films became fully degraded and lost all structure, compare Fig. 4.7b.

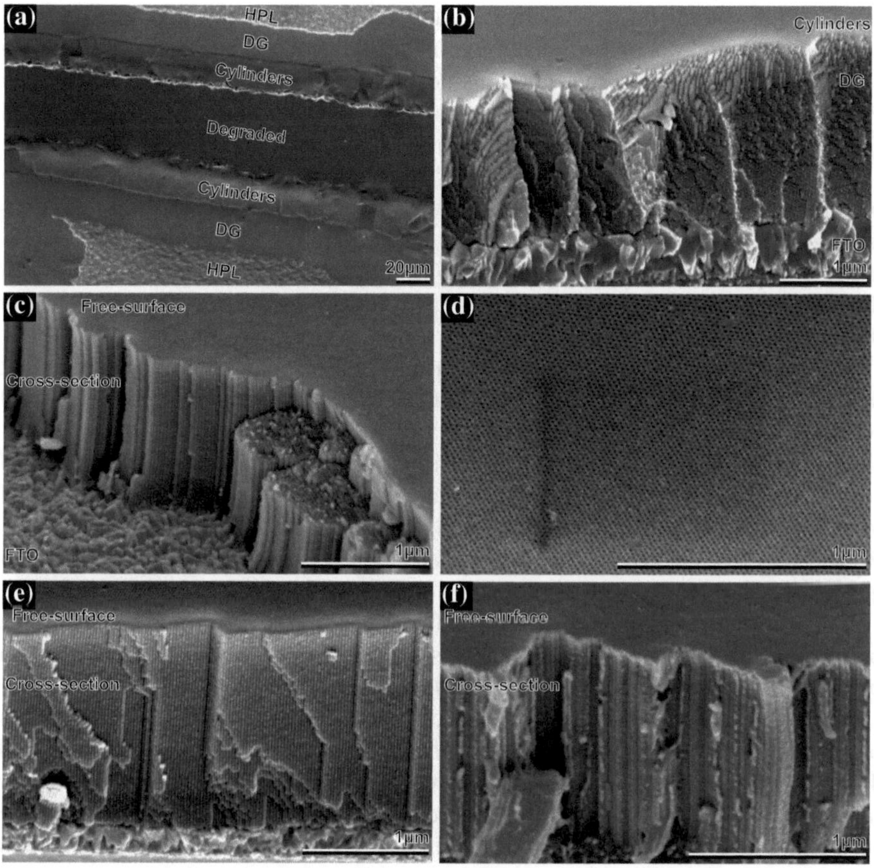

Fig. 4.12 SEM images of PFS18-*b*-PLA76$|_{15\,h}^{170\,°C}$. **a** View onto a film showing HPL, double-gyroid, standing cylinders, and degraded regions. **b** Cross-section of the film showing the interface between double-gyroid and cylinders. The transition from gyroid to standing cylinders starts a the free-surface. **c** 45°—View onto a cross-section of standing cylinders. **d** Free-surface of hexagonally arranged cylinders. **e** Cross-sectional view of cylinders perfectly aligned perpendicular to the FTO-coated substrate. **f** After advancing annealing time the nanostructure shows first signs of polymer degradation

4.3.4 Optimized Brief Thermal Annealing

The results from the prolonged annealing attempts were somewhat misleading, since they suggested that a PLA volume fraction of 41–45 % yields double-gyroids. Nevertheless, they also demonstrated that prolonged annealing is not a feasible option for the materials polymerized in this study. After synthesizing more copolymers as well as testing even more annealing protocols, it was discovered that PFS26-*b*-PLA118 with a PLA volume fraction of 38 % formed a double-gyroid-structured film after just an hour of annealing at around 160 °C. These findings also supported the hypothesis that the PLA volume fraction changes during long thermal annealing due to decomposition and thereby, a phase transition is induced. In contrast, for thermal annealing performed below 180 °C and durations not exceeding an hour, it is believed, that PLA degradation can be neglected and the observed morphology change is a real thermodynamically driven phase transition.

Thereafter, the annealing time was further reduced to 20 min, which proved to be sufficient to form completely double-gyroid-structured films using polymers containing a PXS-OH molecular weight M_n in the range of 11–19 kg mol^{-1}. Importantly, copolymers consisting of PFS, P(F)S, and solely PS all were found to form double-gyroids, but the PClS and PBrS based copolymers tested did not yield any double-gyroids, see Table B.1.

All copolymer which formed DGs during brief annealing had in common that they form a worm-like structure during spin-coating. In contrast the earlier discussed polymers formed the meta-stable HPL phase. The temperature induced phase transition from the worm-like to double-gyroid morphology started to occur at around 160 °C, see Fig. 4.13. The optimal temperature for the studied polymers was found to be between 170 and 180 °C, with the higher molecular weight polymers, such as P(F)S39-*b*-PLA160, P(F)S40-*b*-PLA161, and P(F)S41-*b*-PLA162 requiring temperatures up to 210 °C.

The film thickness did not seem to influence the phase transition, enabling the preparation of templates with thicknesses ranging from 500 nm up to 20 μm. An obvious difference to the double-gyroids formed during prolonged annealing is the size of the self-assembled domains. Typically, the briefly annealed double-gyroids presented domains with dimensions in the micrometer range, see Figs. 4.13 and 4.16. For most industrial applications, however, the order of the morphology is less important, while a short annealing procedure is essential to save cost and enable scalability. As became obvious during template refilling, the grain boundaries did not have a negative effect on the pore connectivity nor the pore diameter.

Not only did the double-gyroid template preparation via the brief thermal annealing prove to be highly robust and reproducible, but so did the PLA copolymerization. Almost all of the last synthesized copolymers, namely PXS-*b*-PLA201 to 215, were successful, yielding up to 2 g of DG-forming material per batch, see Table B.1.

Usually it was possible to tell by eye whether cylinder or DG morphologies were adopted during thermal annealing. Films with a cylindrical structure exhibited undulating free-surfaces with a micron-scale wavelength, which resulted in a milky appearance of the film due to light scattering. In contrast, DG-structured films had a locally very uniform film thickness with a smooth free-surface and were as a result highly transparent.

Increasing the annealing temperature for polymers, which showed the onset of a worm-like to DG transition to around 160 °C, led to another phase transition from the DG phase to cylinders which were aligned perpendicular to the substrate. This presented a possibility to gain more long-range ordered double-gyroid films, as depicted in Fig. 4.14. Firstly, long-range order was established in films of PFS45-*b*-PLA184 by ramping the temperature up from 150 to 205 °C in 8 min and tempering for 3 min during which the copolymers formed aligned cylinders. Then the samples were cooled

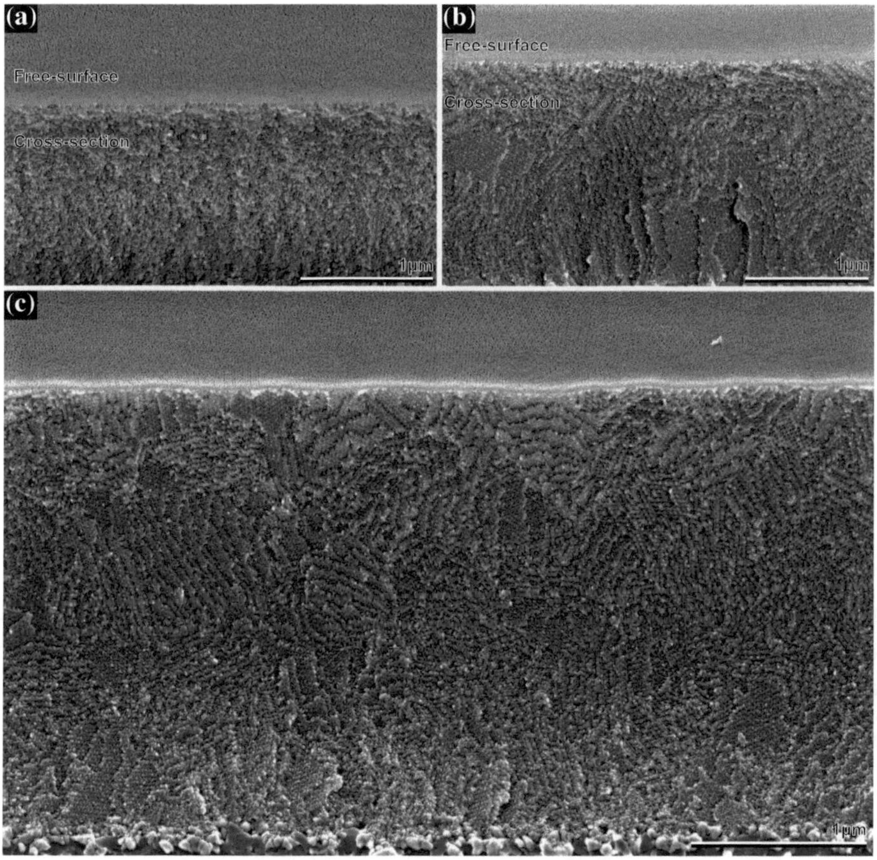

Fig. 4.13 Representative SEM images showing the voided DG films used as templates for functional materials in this thesis. **a**, Cross-section of P(F)S49-*b*-PLA192$|_{0.33\,h}^{150\,°C}$ with a worm-like structure which was also present in as-spun films. **b**, **c**, Cross-section of P(F)S49-*b*-PLA192$|_{0.33\,h}^{174\,°C}$ with DG morphology

to 180 °C within 4 min and were annealed a this temperature for a further 10 min, before quenching to room temperature. During the final annealing duration the DG morphology with long-range order was adopted, but domain dimensions as large as in the case of prolonged annealing were not achieved.

In addition to the various copolymers also a mixture of two copolymers, PFS26-*b*-PLA120:PFS26-*b*-PLA126 = 1:1, as well as a binary copolymer-homopolymer blend, PFS26-*b*-PLA120:PFS29 = 19:1, produced DG-structured films during the 'standard annealing' protocol at 173 °C for 20 min. For industrial application this poses a possibility to compensate for block composition fluctuation that might occur for large-scale copolymerizations.

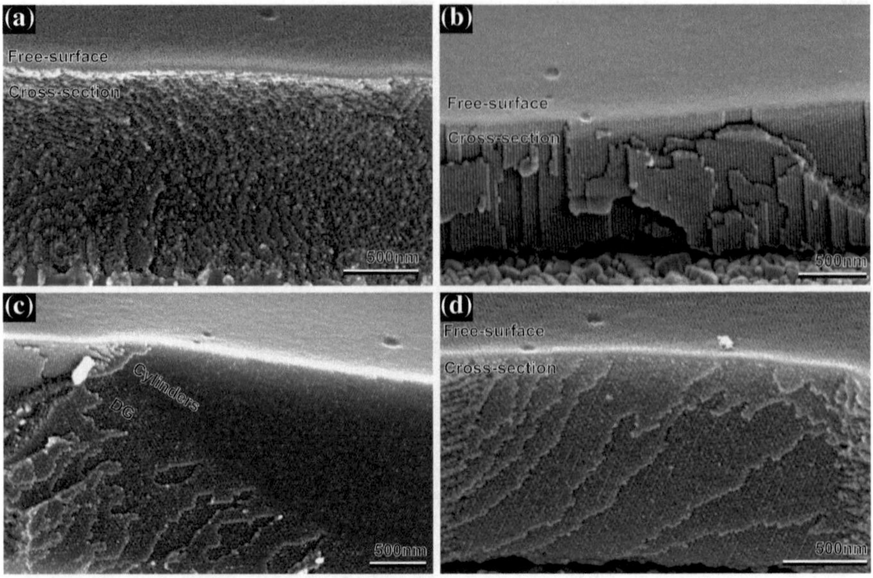

Fig. 4.14 Films of PFS45-*b*-PLA184 prepared under different annealing conditions. **a** PFS45-*b*-PLA184$|_{0.33\,h}^{174\,°C}$ annealed according to the 'standard annealing' procedure. **b** Cylinders orientated perpendicular to the substrate in PFS45-*b*-PLA184$|_{0.25\,h}^{205°C}$. **c**, **d** Phase transition from cylinders to DG occurring as predicted in Fig. 4.5 and ordered DG structure observed in films of PFS45-*b*-PLA184 first annealed at 205 °C and subsequently, at 180 °C, as described in the main text

4.3.5 Microphase Separation Behavior at the Free-Surface

The potential use of voided DG films as organic templates for atomic layer or chemical vapor deposition require porosity at the free-surface. Films of PS50-*b*-PLA214 were frequently covered with a layer of lying cylinders with thicknesses typically not exceeding one unit cell. In contrast, the free-surface of P(F)S49-*b*-PLA192 showed the characteristic DG patterns with the voided PLA network extending across the interface, see Figs. 4.16 and 4.17. Since some of the P(F)S containing copolymer also showed this phase separation behavior at the free-surface interface, I suspect impurities or a non-ideal copolymer composition rather than different surface tensions to be responsible for this effect.

In PS50-*b*-PLA214 films always a few scattered PLA channels were present at the grain boundaries formed by domains of aligned lying cylinders piercing through the otherwise compact free-surface, see Fig. 4.15c. These rare channels were sufficient to allow the infiltration of the underlying DG-structured bulk material of the film with the alkaline etching solution. Further, the infiltration with electrolytes for electrochemical deposition is possible via the few pores. This demonstrates the advantage of a multicontinuous three-dimensional nanostructure. A 1 min oxygen plasma treatment proved to be sufficient to fully remove the semi-porous capping layer.

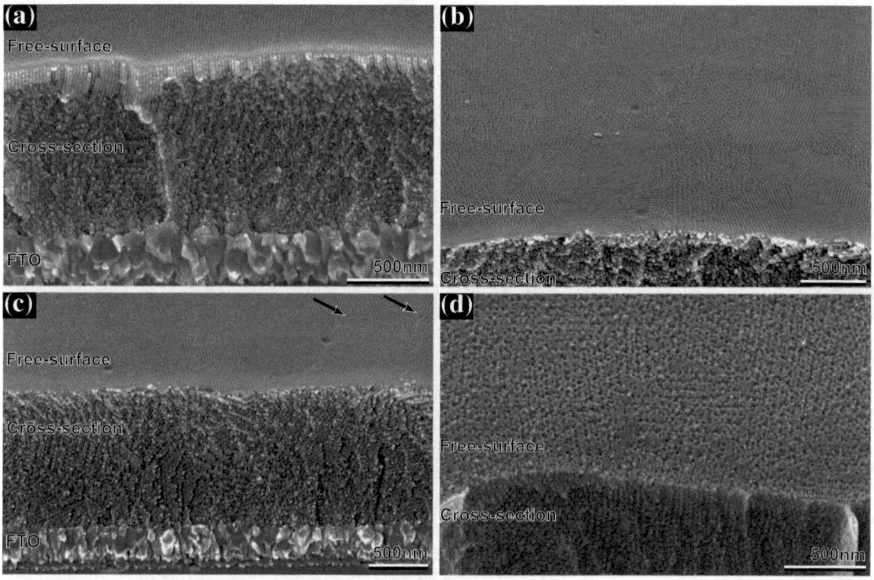

Fig. 4.15 Structural behavior at the free-surface with DG present in the bulk of the film. **a** Rarely observed standing cylinders at the free-surface of PFS26-*b*-PLA118$|_{1\,\mathrm{h}}^{170\,°\mathrm{C}}$1170. **b** Coexistence of the gyroidal double wave pattern and lying cylinders at the free-surface of P(F)S41-*b*-PLA167$|_{0.16\,\mathrm{h}}^{205\,°\mathrm{C}}$. **c** PS50-*b*-PLA214$|_{0.33\,\mathrm{h}}^{173\,°\mathrm{C}}$ film covered with a layer of lying cylinders. *Arrows* are pointing at scattered pores reaching the free-surface. **d** PS50-*b*-PLA214$|_{0.33\,\mathrm{h}}^{173\,°\mathrm{C}}$ film after oxygen plasma treatment for 1 min

The DG unit cell dimension of P(F)S49-*b*-PLA192 were determined by analyzing the SEM images of the double wave patterned free-surface by overlaying a simulated pattern of the (211) plane. The edge length of the simulated square pattern of Fig. 4.16a is 7.1 unit cells or 296 nm. Thus, the unit cell as determined by SEM is 41.7 ± 2 nm, but the error was estimated to be at least 2 nm since the SEM specimen drifted during image acquisition due to charging. A more sophisticated and accurate way to determine the unit cell dimensions were performed by atomic force microscopy (AFM), measuring a value of 43.7 ± 0.2 nm, see Fig. 4.17.

4.3.6 Microphase Separation Behavior at the Film/Substrate Interface

A crucial requirement for the successful replication of the polymer scaffold by electroplating is for the mesopores to extend across both film interfaces to enable electrolyte infiltration. While a possible nonporous layer at the free-surface can be etched subsequent to template fabrication, this is not a feasible option for a wetting layer

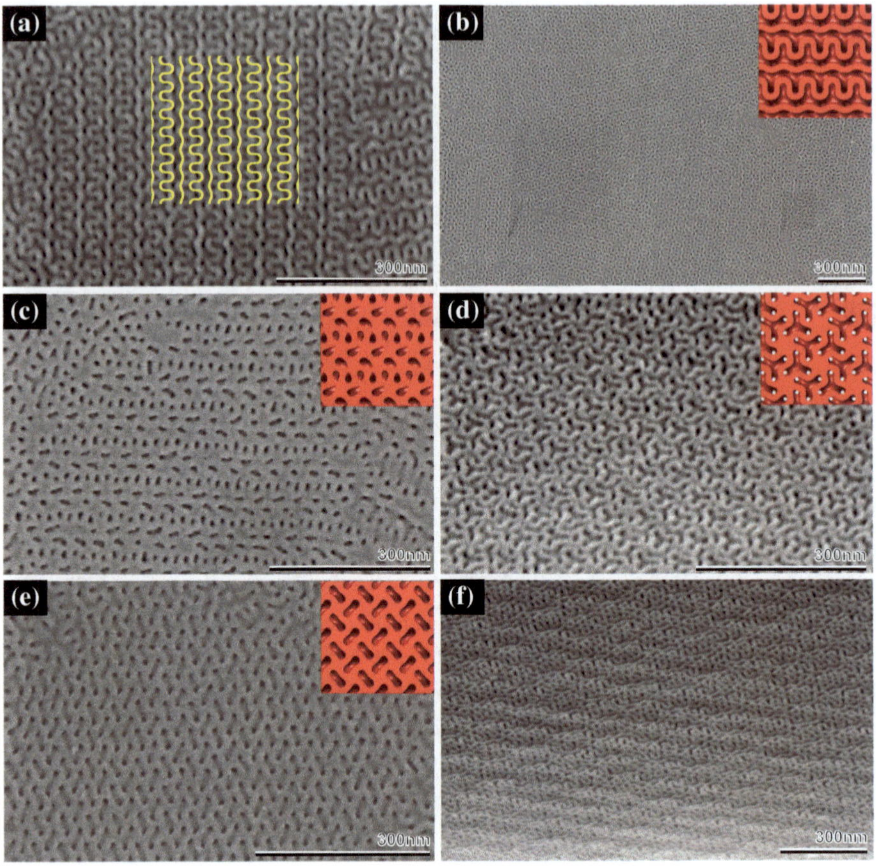

Fig. 4.16 SEM images of the DG-structured free-surface of P(F)S49-*b*-PLA192$|_{0.33\,h}^{174\,°C}$ films. **a** Characteristic double wave pattern of the (211) plane which was identified as cleavage plane, see Fig. 2.5. The simulated pattern is 7.1 × 7.1 unit cells wide or 296 × 296 nm. **b** Differently orientated crystalline DG domains showing the (211) plane result in the formation of grain boundaries. **c** (211) plane. **d** (111) plane. **e** (100) plane. **f** No plane could be assigned to this pattern

formed at the template/substrate interface. Thus, prevention of preferential wetting of the substrate by either of the two polymer chains is essential as well as circumvention of unfavorable phase morphology formation due to surface replication. Crossland et al. prepared neutral FTO-coated substrate surfaces for MHI41 by oxygen plasma cleaning. Unfortunately, this surface treatment proved to be unsuitable for all copolymers synthesized in this study regardless of the PXS block and caused the polymer films to peel off the substrates during PLA etching. An unlikely but possible explanation for these findings might be the interactions between the different end-groups of the polymers and the substrate surface.

Using P(F)S49-*b*-PLA192, various different FTO surface modifications were tested, including piranha etch, RCA clean, ultraviolet and plasma etch in combination

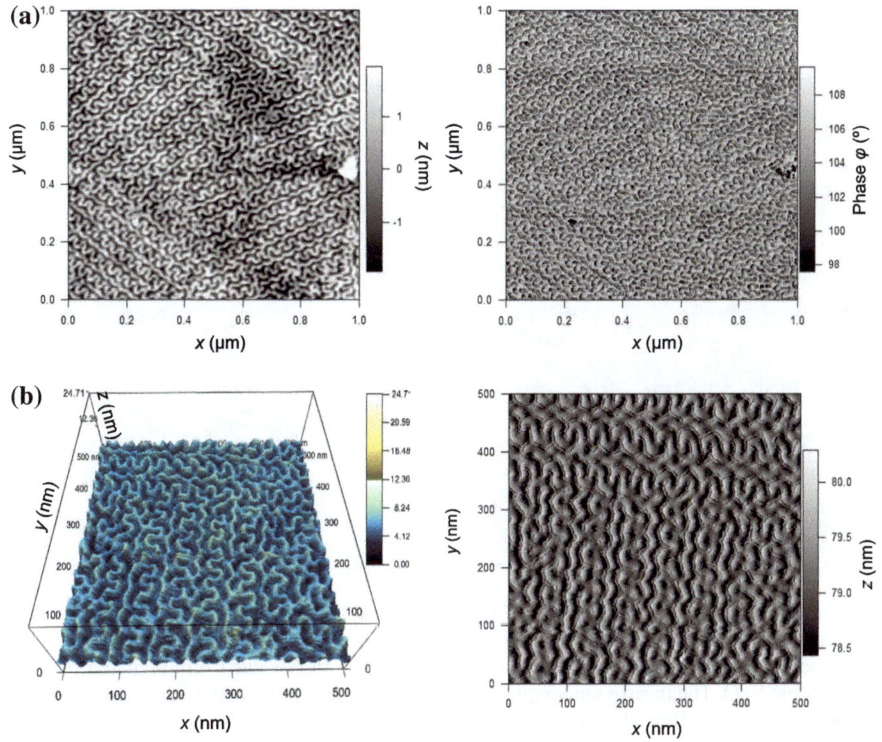

Fig. 4.17 AFM images of the free-surface of P(F)S49-*b*-PLA192 (standard annealing) after PLA removal. The images were taken by Dr. Chris Bower on a MFP-3D Asylum. **a** Height and phase image showing a single double-gyroid domain with the double wave pattern of the (211) plane. An unit cell size of 43.7 ± 0.2 nm was extracted from this data. **b** Corresponding three-dimensional height images

with different silanes such as MTS, OTS, and HMDS. Initially, the lack of precise control over the silanization did not yield satisfying results. Nonetheless, these tests revealed that treatments tailoring hydrophilic FTO surfaces resulted in the release of film from the substrate during PLA etching, suggesting the presence of a PLA wetting layer formed at the substrate surface during film deposition and annealing. In contrast, a nonporous P(F)S wetting layer was formed on substrates with water contact angles well above 60°, see Fig. 4.18.

The degree of silanization and thereby the water wettability is controlled by the exposure time, the amount of silane used, and importantly the concentration of water. Water has a big influence on the mechanism of the silane monolayer formation and structural properties. In the absence of water submonolayers with only one siloxane bond binding can be formed. Water promotes the hydrolysis of the remaining SiCl groups on the initial immobilized silane layer enabling another silanization reaction yielding enhanced silane density. However excess water leads to uncontrolled silane

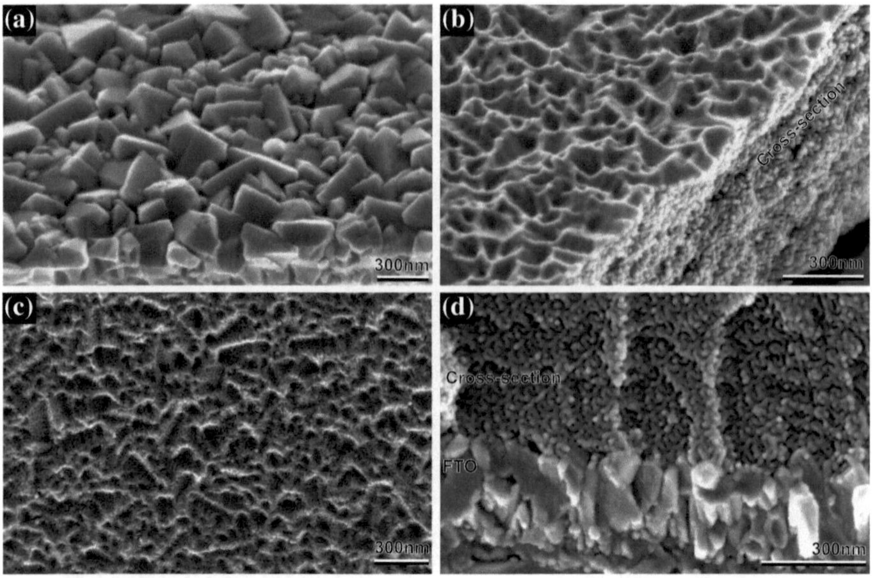

Fig. 4.18 Phase behavior of P(F)S49-*b*-PLA192$|_{0.33\,h}^{174\,°C}$ at the film/substrate interface studied by SEM. **a** SEM of the free-surface of the used FTO-coated glass substrates revealing a rough surface topography. **b** Nonporous film/substrate interface of a PLA etched film prepared on super hydrophobic FTO. The inverse topography of the FTO and the porous cross-section are clearly visible. **c** Highly porous film/substrate interface of a PLA etched film prepared on a FTO substrate, the surface of which was modified according to Sect. 4.2.2. **d** Cross-sectional view of the film shown in (**c**) revealing that the DG phase starts directly at the interface with no other phase such as cylinders or HPL present

polymerization with multilayer formation. Silanization in a water free environment such as a glove box was too slow, while gas phase deposition in an evacuated desiccator and from a toluene solution generally resulted in super hydrophobic surfaces.

After a tedious optimization process, silanization with OTS dissolved in anhydrous cyclohexane in air was found to provide a neutral surface for P(F)S49-*b*-PLA192 when controlling the amount of water dissolved in the silanization solution and absorbed on the substrate surface. It is believed that this optimized treatment yields a partial coverage with OTS giving rise to a water contact angle of $59 \pm 5°$. The necessary hydroxyl groups on the FTO-surface were introduced by cleaning the substrates in a piranha solution. The subsequent complete removal of the piranha solution is essential, otherwise it leads to degradation of the polymer during thermal annealing. A piranha clean was preferred to plasma clean, due to issues with redeposition of the etched material from the plasma chamber onto the substrate. To quantify the degree of porosity at the film/substrate interface the polymer films were peeled off the substrates using carbon tape and attached to SEM stubs for imaging, see Sect. 4.2.2.

Since unsubstituted PS is far less hydrophobic than PFS or P(F)S a change of the styrenic block composition alters the copolymer's interfacial and wetting behav-

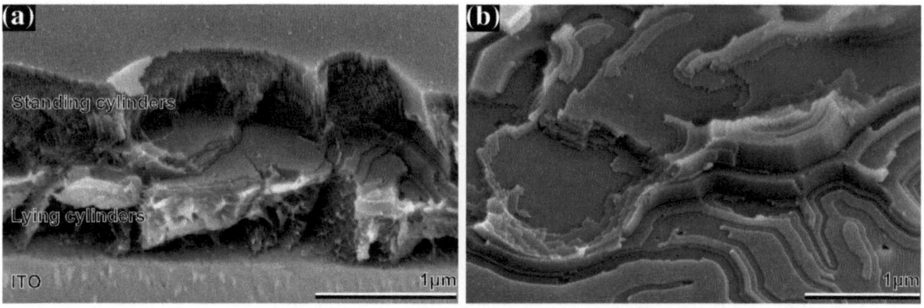

Fig. 4.19 Phase behavior of P(F)S49-*b*-PLA192$|^{174\,°C}_{0.33\,h}$ on silanized ITO-coated glass. **a** Cross-section revealing that the *top* part of the film formed cylinders orientated perpendicular to the substrate which were voided during PLA etch. In contrast, cylinders aligned parallel to the substrate were adopted at the *bottom* part of the film which were not voided because of lacking etchant infiltration. Compared to Fig. 4.15 this highlights the advantages of a multicontinuous structure like the DG. **b** *Top-view* onto lying cylinders forming a terrace-like structure

ior. Thus, it was not obvious that the substrate surface treatment optimized for P(F)S49-*b*-PLA192 would be applicable for PFS or PS containing copolymers. Fortunately, the surface modification provided a neutral surface for all synthesized DG-forming copolymers. On silane treated rough FTO-coated substrates the DG morphology started directly at the interface and no other phase such as cylinders or HPL were observed, see Fig. 4.18.

On smooth ITO-coated substrates this surface treatment involving an oxygen plasma clean instead of piranha etch turned out to be unsuccessful, see Fig. 4.19 [54]. Cylinders aligned parallel to the substrate surface seemed to be energetically favored. In contrast, films prepared on fairly smooth gold-coated glass or silicon did not show this behavior but adopted the DG structure. Further, flexible and temperature resistant poly(ethylene naphthalate) (PEN) foils are currently tested as substrates in collaboration with Dr. Chris Bower [55]. PEN foils roughened by an oxygen plasma and coated with a semi-transparent, conducting gold layer were successfully tested as substrates.

While the silanization treatment was performed to achieve a porous film/substrate interface required for electrochemical deposition, an insulating layer between the template and the conducting FTO coating can be utilized to pattern the deposition. Visible design patterns were manufactured by further modifying the silanized FTO surface using standard photo-lithography to create an isolating SU-8 pattern, which locally inhibited electroplating [49].

4.3.7 Bulk Phase Behavior

The temperature dependent bulk phase behavior was further study by small angle X-ray scattering (SAXS) at the Diamond synchrotron for bulk samples of P(F)S49-*b*-

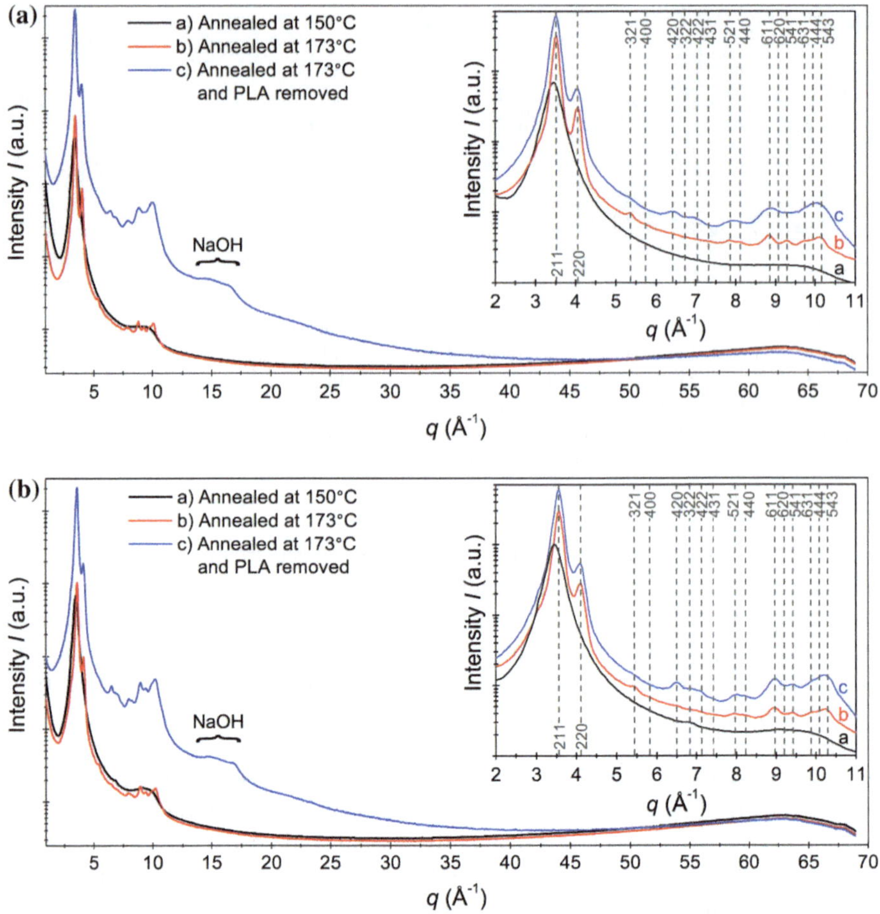

Fig. 4.20 SAXS pattern for bulk samples of **a** P(F)S49-*b*-PLA192 and **b** PS50-*b*-PLA214 annealed at 150 °C (*a*) and 173 °C (*b*). For the voided styrenic scaffold the scattered intensity for small wave vectors q increased significantly due to enhanced contrast (*c*). Curve *a* corresponds to a disordered worm-like morphology as confirmed by SEM. Peaks with scattering wave vector spacing-ratios consistent with a $Ia\bar{3}d$ gyroid phase are discernible in case of sample *b* and *c*. The peaks of curve *c* located in the q range from 13 to 18 Å^{-1} can be attributed to remaining PLA etchant, sodium hydroxide

PLA192 and PS50-*b*-PLA214, see Fig. 4.20. Temperature annealing was performed according to the thin film annealing protocol at 150 and 173 °C. Further, voided poly(styrene) scaffolds were studied, which were obtained by selective etching of the PLA phase with sodium hydroxide according to Sect. 4.2.5 for 48 h. The peaks observed for samples annealed at 173 °C had scattering wave vector spacing-ratios of $\sqrt{6}$, $\sqrt{8}$, $\sqrt{14}$, $\sqrt{16}$, $\sqrt{20}$, $\sqrt{22}$, $\sqrt{24}$, $\sqrt{26}$, $\sqrt{30}$, $\sqrt{32}$, $\sqrt{38}$, $\sqrt{40}$, $\sqrt{42}$, $\sqrt{46}$, $\sqrt{48}$, and $\sqrt{50}$, consistent with a $Ia\bar{3}d$ gyroid phase. The cubic unit cell length a is given by

$$a = \frac{2\pi}{q}\sqrt{h^2 + k^2 + l^2}, \tag{4.8}$$

where q is the reciprocal lattice vector [56]. For P(F)S49-*b*-PLA192 and PS50-*b*-PLA214 values of 43.7 and 43.2 nm were measured, respectively. The determined unit cell size of P(F)S49-*b*-PLA19 agrees with the value measured by atomic force microscopy.

4.3.8 Surface Area Measurement

The internal surface area of the double-gyroid structure formed by P(F)S49-*b*-PLA192 was indirectly measured by hydrogen adsorption using cyclic voltammetry [44, 57, 58]. BET surface area measurements were not performed since larger quantities of the precious material would have been needed. For this purpose a platinum replica of the double-gyroid with a known deposition area (0.95 mm^2) and thickness (1.15 μm) was prepared by electroplating on a SU-8 patterned FTO substrate, see Fig. 4.21 [49]. The templated electrodeposition of platinum is discussed in detail in Chap. 6.

The surface area of the platinum array was determined by measuring the charge associated with hydrogen adsorption in a 0.1 M sulfuric acid (99.999 %) aqueous solution, which was deoxygenated by bubbling with nitrogen for a minimum of 10 min. The samples were cleaned by cycling between −0.3 and 1.36 V versus Ag/AgCl for at least fifty cycles before recording the stabilized cyclic voltammogram presented in Fig. 4.21e. The measured hydrogen absorption charge per bulk volume was $-338.8\,\mathrm{C\,cm^{-3}}$. With a conversion factor of $210\,\mu\mathrm{C\,cm^{-2}}$ this translates to an electrochemically accessible surface area to bulk volume ratio of $161.4\,\mu\mathrm{m}^{-1}$ or $7.05 \cdot L^{-1}$ for the corresponding unit cell dimensions of $L = 43.7$ nm, see Sect. 4.3.7. For a PLA volume fraction $f_{\mathrm{PLA}}^{\mathrm{NMR}} = 37.9\,\%$, which was determined by NMR (see Table 3.6), the computer simulation predicts a specific surface area per bulk volume of $4.94 \cdot L^{-1}$ as depicted in Fig. 2.4. The discrepancy between the measured and predicted surface areas is indicative of the presence of sub-nanometer surface roughness, which contributes to the overall surface area.

4.3.9 Nanostructured Colloids

Combining micro- and macrophase separation allows to prepare nanostructured colloids via good-solvent evaporation [59, 60]. Although refilling with functional materials might prove to be challenging, numerous possible applications for these nanoporous particles have been identified, including drug delivery, filters, and catalysts with high surface areas. Only preliminary results were obtained during this study, thus a brief experimental description is given here. Water was very slowly added to a stirring 1 % solution of PFS27-*b*-PLA128 in tetrahydrofuran (water and

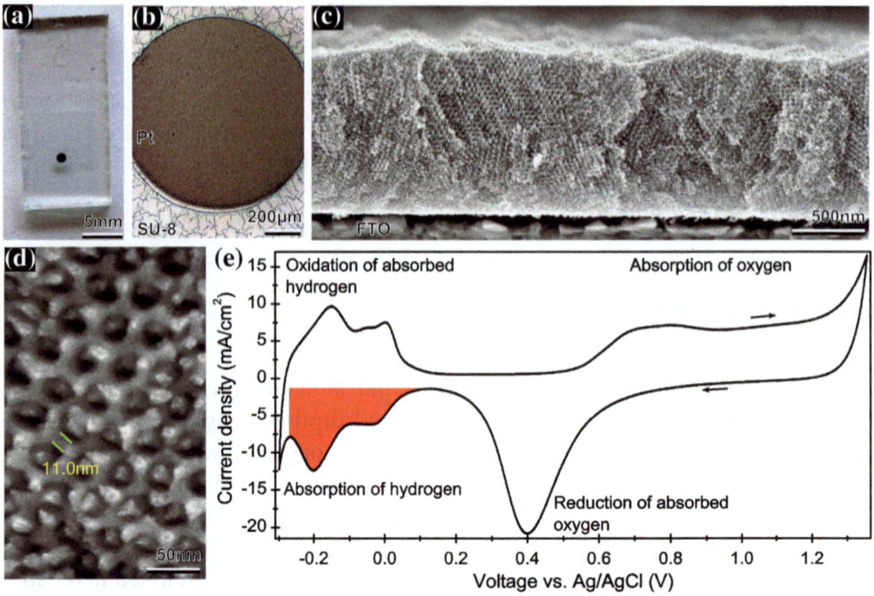

Fig. 4.21 Electrochemical surface area measurement using hydrogen adsorption. **a** Photograph of black, circularly shaped platinum on a FTO substrate after dissolution of the polymer template. **b** Micrograph of the replicated platinum with an area of 0.95 mm^2, revealing smooth and homogenous deposition. **c** Cross-sectional SEM image of the platinum double-gyroid with a film thickness of 1.15 μm. **d** High-resolution SEM of the platinum double-gyroid showing the (110) plane. **e** Cyclic voltammogram recorded for a gyroid-structured platinum electrode in 0.1 M sulfuric acid with a potential scan rate of 50 mV s^{-1}. The red area denotes the hydrogen absorption charge density of -38.97 mC cm^{-2}

THF are miscible). Then the good-solvent THF which has a higher vapor pressure than water was allowed to slowly evaporate. Due to the fact that neither of the two copolymer blocks is water soluble, droplets with a high polymer and THF concentration form which finally form colloids when all the THF has evaporated. Then, the PLA phase was selectively removed by the addition of NaOH and methanol. The nanostructured particles were then collected by centrifugation, dispersed on a substrate, and characterized by means of SEM, see Fig. 4.22. Adjusting the polymer concentration and the THF:water ratio allows to control the formed particle size.

The involved solvent might preferentially swell one of the copolymer block and thereby, influence the thermodynamically stable morphology adopted during microphase separation. Thermal annealing might be necessary to achieve the desired morphology. A possible approach would be to disperse the colloids in an alkane-solvent, which are 'non-solvents' for PXS-*b*-PLA copolymers, and perform an annealing step using an autoclave.

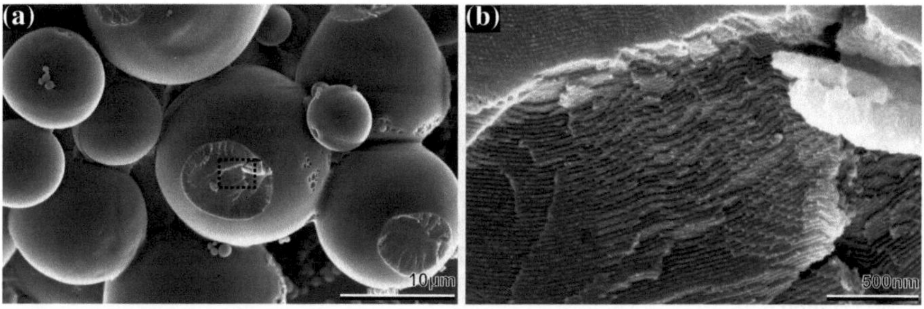

Fig. 4.22 **a** Nanostructured styrenic colloids prepared by good-solvent evaporation. **b** Magnification of (**a**) revealing a cylindrical microphase separated structure

4.4 Concluding Remarks

In summary, a protocol for the reliable, time and energy efficient preparation of fully porous double-gyroid thin films on various substrates for the further usage as templates for functional materials was established. This includes a surface modification and a fast thermal annealing protocol to induce a phase transition to the double-gyroid. Compared to the earlier long annealing protocol of MHI41, which took up to several days, the new rapid thermal annealing for PXS-*b*-PLA synthesized by ATRP and organocatalytic ROP makes the production of voided double-gyroid templates industrially applicable.

Additionally, it was shown that PFS, P(F)S, and PS are all suitable as the styrenic blocks of PXS-*b*-PLA, capable of forming DG-structured thin films as long as the PXS has a molecular weight lower than 20 kg mol^{-1}. Importantly, it was shown that the styrenic block does not have to contain any fluorinated monomer, which makes the template production significantly cheaper and environmentally sustainable. Further, the copolymer synthesis on a 2 g scale was robust and highly repeatable and yielded DG-forming material. Unfortunately, the copolymers containing ultraviolet cross-linkable 4-bromostyrene and 4-chlorostyrene did not form the desired DG [61–63]. However, only a few different copolymer compositions were tested and it is highly likely that copolymers containing these monomers are able to form the DG structure.

The brief thermal annealing resulted in DG films consisting of small ordered self-assembled domains. Some applications might however require longer ranged order. A possible way to improve the domain size of the copolymer films might be the application an electric field during an initial annealing stage which favors the formation of an aligned cylindrical phase orientated perpendicular to the substrate [46]. Subsequently, switching-off the electrical field should then induce a transition to a double-gyroid morphology with long-range order.

References

1. M.R.J. Scherer, L. Li, P.M.S. Cunha, O.A. Scherman, U. Steiner, Adv. Mater. **24**(9), 1217–1221 (2012)
2. A.S. Zalusky, R. Olayo-Valles, C.J. Taylor, M.A. Hillmyer, J. Am. Chem. Soc. **123**(7), 1519–1520 (2001)
3. H. Mao, M.A. Hillmyer, Soft Matter **2**(1), 57–59 (2006)
4. P.J. Flory, J. Chem. Phys. **10**(1), 51–61 (1942)
5. F.S. Bates, G.H. Fredrickson, Phys. Today **52**(2), 32 (1999)
6. A.S. Zalusky, R. Olayo-Valles, J.H. Wolf, M.A. Hillmyer, J. Am. Chem. Soc. **124**(43), 12761–12773 (2002)
7. M. Rubinstein, R.H. Colby, *Polymer Physics*, (Oxford University Press, 2003)
8. M.W. Matsen, M. Schick, Phys. Rev. Lett. **72**(16), 2660–2663 (1994)
9. M.W. Matsen, F.S. Bates, Macromolecules **29**(4), 1091–1098 (1996)
10. A.K. Khandpur, S. Foerster, F.S. Bates, I.W. Hamley, A.J. Ryan, W. Bras, K. Almdal, K. Mortensen, Macromolecules **28**(26), 8796–8806 (1995)
11. H.-W. Park, K. Im, B. Chung, M. Ree, T. Chang, K. Sawa, H. Jinnai, Macromolecules **40**(7), 2603–2605 (2007)
12. K. Yamada, M. Nonomura, T. Ohta, Macromolecules **37**(15), 5762–5777 (2004)
13. C.A. Tyler, D.C. Morse, Phys. Rev. Lett. **94**(20), 208302 (2005)
14. M.I. Kim, T. Wakada, S. Akasaka, S. Nishitsuji, K. Saijo, H. Hasegawa, K. Ito, M. Takenaka, Macromolecules **42**(14), 5266–5271 (2009)
15. R.-M. Ho, Y.-W. Chiang, C.-K. Chen, H.-W. Wang, H. Hasegawa, S. Akasaka, E.L. Thomas, C. Burger, B.S. Hsiao, J. Am. Chem. Soc. **131**(51), 18533–18542 (2009)
16. M. Matsen, Phys. J. E **30**(4), 361–369 (2009)
17. M. Takenaka, T. Wakada, S. Akasaka, S. Nishitsuji, K. Saijo, H. Shimizu, M.I. Kim, H. Hasegawa, Macromolecules **40**(13), 4399–4402 (2007)
18. M.I. Kim, T. Wakada, S. Akasaka, S. Nishitsuji, K. Saijo, H. Hasegawa, K. Ito, M. Takenaka, Macromolecules **41**(20), 7667–7670 (2008)
19. N.A. Lynd, A.J. Meuler, M.A. Hillmyer, Prog. Polym. Sci. **33**(9), 875–893 (2008)
20. D.M. Cooke, A.-C. Shi, Macromolecules **39**(19), 6661–6671 (2006)
21. M. W. Matsen, Phys. Rev. Lett. 2007, 99(14), 148304, copyright (C) 2010 The American Physical Society; Please report any problems to prola@aps.org.
22. D. Bendejacq, V. Ponsinet, M. Joanicot, Y.-L. Loo, R.A. Register, Macromolecules **35**(17), 6645–6649 (2002)
23. M. R. J. Scherer. Nanostructured Materials via Self-Assembled Templates, University of Konstanz, 2009
24. N.A. Lynd, M.A. Hillmyer, Macromolecules **38**(21), 8803–8810 (2005)
25. S. Abbas, C.D. Cowman, T.P. Lodge, Macromol. Rapid Commun. **30**(4–5), 352–361 (2009)
26. C. Luo, W. Huang, Y. Han, Macromol. Rapid Commun. **30**(7), 515–520 (2009)
27. M.W. Matsen, Phys. Rev. Lett. **80**(20), 4470–4473 (1998)
28. M. Ima, K. Sakai, M. Kikuchi, K. Nakaya, A. Saeki, T. Teramoto, J. Chem. Phys. **122**(21), 214906–214906-10 (2005)
29. X. Cheng, L. Lin, W.E.P. Zhang, A.-C. Shi, Phys. Rev. Lett. **104**(14), 148301 (2010)
30. D. Cam, M. Marucci, Polymer **38**(8), 1879–1884 (1997)
31. H. Abe, N. Takahashi, K.J. Kim, M. Mochizuki, Y. Doi, Biomacromolecules **5**(4), 1606–1614 (2004)
32. X. Liu, Y. Zou, W. Li, G. Cao, W. Chen, Polym. Degrad. Stab. **91**(12), 3259–3265 (2006)
33. M.C. Dalvi, T.P. Lodge, Macromolecules **26**(4), 859–861 (1993)
34. C. Shin, D.Y. Ryu, J. Huh, J.H. Kim, K.-W. Kim, Macromolecules **42**(6), 2157–2160 (2009)
35. I. Podariu, A. Chakrabarti, J. Chem. Phys. **118**(24), 11249–11257 (2003)
36. Y. Yin, P. Sun, R. Jiang, B. Li, T. Chen, Q. Jin, D. Ding, A.-C. Shi, J. Chem. Phys. **124**(18), 184708–184708 (2006)

37. B. Lee, I. Park, J. Yoon, S. Park, J. Kim, K.-W. Kim, T. Chang, M. Ree, Macromolecules **38**(10), 4311–4323 (2005)
38. I. Park, B. Lee, J. Ryu, K. Im, J. Yoon, M. Ree, T. Chang, Macromolecules **38**(25), 10532–10536 (2005)
39. H.-W. Park, J. Jung, T. Chang, K. Matsunaga, H. Jinnai, J. Am. Chem. Soc. **131**(1), 46–47 (2009)
40. A.S. Finnemore, M.R.J. Scherer, R. Langford, S. Mahajan, S. Ludwigs, F.C. Meldrum, U. Steiner, Adv. Mater. **21**(38–39), 3928–3932 (2009)
41. T.H. Epps, D.M. DeLongchamp, M.J. Fasolka, D.A. Fischer, E.L. Jablonski, Langmuir **23**(6), 3355–3362 (2007)
42. E.J.W. Crossland, S. Ludwigs, M.A. Hillmyer, U. Steiner, Soft Matter **3**(1), 94–98 (2007)
43. E. J. W. Crossland. Block Copolymer Patterning of Functional Materials, University of Cambridge, 2008
44. E.J.W. Crossland, M. Kamperman, M. Nedelcu, C. Ducati, U. Wiesner, D.M. Smilgies, G.E.S. Toombes, M.A. Hillmyer, S. Ludwigs, U. Steiner, H.J. Snaith, Nano Lett. **9**(8), 2807–2812 (2009)
45. E.J.W. Crossland, M. Nedelcu, C. Ducati, S. Ludwigs, M.A. Hillmyer, U. Steiner, H.J. Snaith, Nano Lett. **9**(8), 2813–2819 (2009)
46. E.J.W. Crossland, S. Ludwigs, M.A. Hillmyer, U. Steiner, Soft Matter **6**(3), 670–676 (2010)
47. A.J. Meuler, C.J. Ellison, J. Qin, C.M. Evans, M.A. Hillmyer, F.S. Bates, J. Chem. Phys. **130**(23), 234903–234917 (2009)
48. S.R. Wasserman, Y.T. Tao, G.M. Whitesides, Langmuir **5**(4), 1074–1087 (1989)
49. H. Lorenz, M. Despont, N. Fahrni, N. LaBianca, P. Renaud, P. Vettiger, J. Micromech. Microeng. **7**(3), 121–124 (1997)
50. E.J.W. Crossland, P. Cunha, S. Ludwigs, M.A. Hillmyer, U. Steiner, ACS Appl. Mater. Interfaces **3**(5), 1375–1379 (2011)
51. H. Yu, N. Huang, C. Wang, Z. Tang, J. Appl. Polym. Sci. **88**(11), 2557–2562 (2003)
52. Y. Fan, H. Nishida, Y. Shirai, T. Endo, Polym. Degrad. Stab. **84**(1), 143–149 (2004)
53. H.-Y. Hsueh, Y.-C. Huang, R.-M. Ho, C.-H. Lai, T. Makida, H. Hasegawa, Adv. Mater. **23**(27), 3041–3046 (2011)
54. G. Brown, A. Chakrabarti, J. Chem. Phys. **101**(4), 3310–3317 (1994)
55. V. Zardetto, T.M. Brown, A. Reale, A. Di Carlo, J. Polym. Sci., Part B: Polym. Phys. **49**(9), 638–648 (2011)
56. C. Leal, N.F. Bouxsein, K.K. Ewert, C.R. Safinya, J. Am. Chem. Soc. **132**(47), 16841–16847 (2010)
57. S. Trasatti, O. Petrii, J. Electroanal. Chem. **327**(1–2), 353–376 (1992)
58. J.M. Doña Rodríguez, J.A. Herrera Melián, J. Pérez Peña, J. Chem. Educ. **77**(9), 1195–1197 (2000)
59. H. Yabu, T. Higuchi, K. Ijiro, M. Shimomura, Chaos **15**(4), 047505 (2005)
60. T. Higuchi, A. Tajima, K. Motoyoshi, H. Yabu, M. Shimomura, Angew. Chem. Int. Ed. **48**(28), 5125–5128 (2009)
61. S. Imamura, J. Electrochem. Soc. **126**(9), 1628–1630 (1979)
62. A. Andrady, *In Physical Properties of Polymers Handbook*, 2nd edn. (Springer, Berlin, 2007)
63. Q. Lin, *In Physical Properties of Polymers Handbook* (Springer, Berlin, 2007)

Chapter 5
Templating of Metal Oxides by Electrodeposition

The aim of this part of the study was to use the voided DG films, typically prepared from P(F)S49-*b*-PLA192$|^{174\,^\circ\mathrm{C}}_{0.33\,\mathrm{h}}$ and PS50-*b*-PLA214$|^{173\,^\circ\mathrm{C}}_{0.33\,\mathrm{h}}$, for the template-assisted electrodeposition of nanostructured transition metal oxide (MO) or metal hydroxide thin films. MOs are commonly utilized for their catalytic activity, intercalation and semi-conductive properties. Nanostructuring these MOs is believed to enhanced their unique material properties by providing a large active area and simultaneous improving the access to the bulk material. Subsequent to the deposition of the MO films, the selective removal of the organic scaffold by dissolution yields the desired mesoporous MO network, which is characterized by a high surface area to bulk volume ratio, ideal for many nanotechnology applications, see Fig. 5.1.

A first successful application of a bicontinuous metal-oxide nanostructure based on the DG morphology, obtained by the templated electrochemical deposition of hydrated titanium oxide, has been demonstrated in dye-sensitized solar cells [3]. In contrast to this earlier example, a functional device based on the DG morphology that out-performs existing technologies has yet to be demonstrated. Here, the effects of a three-dimensionally ordered nanostructure on the lithium ion intercalation properties of vanadium pentoxide (V_2O_5) are studied and tested by means of their electrochromic and supercapacitor performance [4]. The former study was conducted in collaboration with **Li Li**, and the latter together with **Dr. Di Wei**. The corresponding figures of merit, including the electrochromic switching speed, coloration contrast, coloration efficiency, and the capacitive energy and power density, are determined and evaluated. Furthermore, the charge storage and chromogenic properties are combined in an electrochromic supercapacitor.

M. R. J. Scherer, *Double-Gyroid-Structured Functional Materials*,
Springer Theses, DOI: 10.1007/978-3-319-00354-2_5,

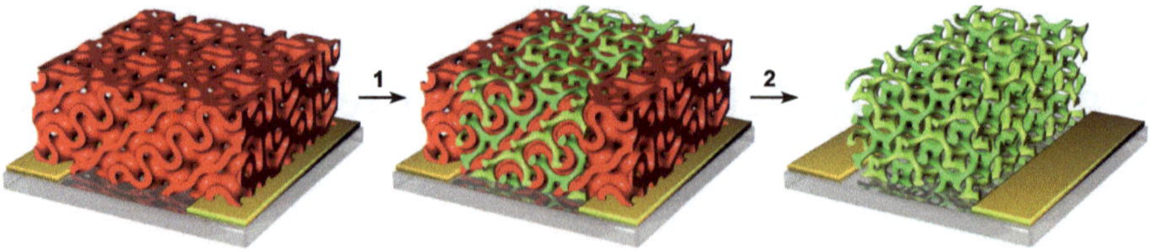

Fig. 5.1 Templated electrodeposition of metal oxides. **(1)** The electrochemical refilling process starts at the FTO surface and progresses trough the voided DG channels. The deposition process is restricted to areas that are not covered by SU-8, thereby creating a visible design pattern in the electroplated V_2O_5. **(2)** Removal of the styrenic template yields the free-standing, mesoporous DG-structured metal oxide network

5.1 Introduction to Electrochemistry

The field of chemistry concerned with the interrelation of electrical and chemical effects, especially the study of chemical changes caused by an electric current and the electrical energy production by chemical reactions, is termed electrochemistry [5]. While electrochemistry encompasses a huge array of different phenomena applied in a variety of technologies, applications, and characterization techniques, such as the surface area measurement by hydrogen adsorption discussed in Sect. 4.3.8, the main emphasis here will be focused on electrodeposition and devices based on electrochemistry, such as electrochemical supercapacitors and electrochromic displays.

5.1.1 Electrochemical Deposition

5.1.1.1 Fundamentals of Electrodeposition

Thin film coatings of various conducting or semiconducting materials, including pure metals and alloys, ceramic oxides, and conjugated polymers, can be prepared on conducting or semiconducting substrates by the low-cost and extremely versatile technique of electrodeposition [6]. The deposition setup consists of a potentiostat and an electrochemical cell holding a liquid electrolyte containing soluble precursors of the material to be deposited in which at least two electrodes are immersed. The substrate to be coated forms the so-called working electrode which is placed opposite and in close proximity to the charge balancing counter electrode. This two-electrode setup is usually used for electrodeposition under galvanostatic conditions, that is, by passing a constant current through the cell. Typically, a third electrode with a stable and well-known electrode potential, such as a calomel or Ag/AgCl reference electrode, is introduced as reference for the working electrode's relative potential measurement. This three-electrode setup is typically employed for potentiostatic deposition where the potentiostat keeps the working electrode's potential stable with respect to the reference electrode. The current-voltage characteristics recorded synchronously by the potentiostat contains valuable information about the deposition mechanism.

In general, electrodeposition is based on the local oxidation or reduction of precursors at the electrode-electrolyte interface driven by the charge q passing trough the electrochemical cell. The redox reaction taking place in close vicinity of the working electrode produces insoluble species that precipitate and form the electrodeposit. Hence, the coating grows preferentially from the conducting substrate and the amount of deposit n, according to Faraday's first law of electrolysis, is given by

$$n = \frac{q}{zF}, \tag{5.1}$$

where F is the Faraday constant and z the number of electrons involved in each redox reaction. This simple relation typically allows a precise control over the deposit's thickness making electrodeposition an extremely useful technique for the refilling of mesoporous thin film templates with a predefined fill fraction. However, the deposition only proceeds where the electrolyte is able to penetrate the pore structure and reach the underlying conducting substrate.

In essence, the electric potential E applied to the working electrode acts as thermodynamic driving force for the electrodeposition which only takes place if E is chosen appropriately with respect to the redox potential E_{redox} of the involved redox couple, given by the Nernst equation [6]

$$E_{\text{redox}} = E^0 + \frac{RT}{zF} \ln \frac{a_{\text{ox}}}{a_{\text{red}}}. \tag{5.2}$$

Here, a_{ox} and a_{red} are the activities of the oxidized and reduced form of the species, respectively, and $E^0 = -\Delta G^0/zF$ is the standard electrode potential of the redox couple A/A^{n+} defined by the Gibbs free energy ΔG^0 of the reaction $A^{n+} + ne^- \rightarrow A$. Thermodynamically, the electrodeposition only proceeds when the relation $E > E_{\text{redox}}$ in case of an oxidation and $E < E_{\text{redox}}$ in case of a reduction is fulfilled [6].

Further, electrochemical kinetic aspects have to be considered, which are described by the Butler-Volmer equation [5]

$$j = j_0 \left[e^{\frac{\alpha zF}{RT}(E-E_0)} - e^{-\frac{(1-\alpha)zF}{RT}(E-E_0)} \right], \tag{5.3}$$

here j is the current density, j_0 the exchange current density, α the symmetry factor and $E - E_0$ the so-called overpotential. For non-zero overpotentials a current flow is observed, but for small j_0 a successful electrodeposition may require a considerably high overpotential. The latter is also necessary to overcome a possible potential barrier hindering the electroactive species to reach the electrode surface and the charge transfer at the electrode surface itself. In order to keep the applied potential low and thereby achieve good control over the deposition and avoid undesired side reactions involving the solvent or dissolved additives, counter electrodes made from high electrocatalytic active materials requiring a small overpotential, such as inert

Pt or a reactive metal electrode including Fe, Co, or Ni depending on the desired product, are usually employed [5].

In addition to the activation overpotential according to Eq. (5.3), the concentration overpotential has to be considered, the contribution of which gains in importance when deposition is performed in the channels of a mesoporous template. This overpotential arises from differences in the concentration c of the electroactive species in close vicinity of the electrode surface compared to the bulk solution. A concentration gradient $\mathrm{d}c/\mathrm{d}x$ is formed when the surface reaction is sufficiently fast and is increased when the flux of electroactive species is hindered by a template. The transport of charge carriers is given by the Nernst-Planck equation

$$j = -D\frac{\mathrm{d}c}{\mathrm{d}x} - cD\frac{zF}{RT}\frac{\mathrm{d}\Phi}{\mathrm{d}x} + cv. \tag{5.4}$$

where v is the flow speed, $\mathrm{d}\Phi/\mathrm{d}x$ the potential gradient, and D the diffusion coefficient.

5.1.1.2 Electrochemical Synthesis of Metal Oxides and Hydroxides

In the following the deposition of semiconducting MO is discussed, while the electroplating of metals and electropolymerization of conjugated polymers are considered in Chaps. 6 and 7, respectively. A variety of metal oxides are electrochemically depositable, such as NiO, V_2O_5, Cu_2O, ZnO, and TiO_2, but this method is limited to conducting or semiconducting oxides [7]. Compared to the conceptually simple cathodic deposition of metals by the reduction of the corresponding cations present in the electrolyte to the elementary metal onto the cathode, MO deposition is more complex. In fact, cathodic and anodic electrodeposition methods for MOs are employed [8].

The most traditional approach for the electrodeposition of metal oxides and hydroxides is based on the cathodic electrogeneration of base [6]. Depending upon the deposition potential, choice of the anion and the pH of the solution, various reactions take place at the cathode, such as the reduction of nitrate ions in aqueous solutions. These reduction reactions cause a local pH increase in vicinity of the cathode either by the consumption of protons or by the generation of hydroxide ions as in the case of nitrate reduction, which subsequently leads to the precipitation of metal hydroxides. The best known application of this technique is the synthesis of nickel hydroxide by the electroreduction of aqueous nickel nitrate solutions. Depending on the metal and deposition conditions, this cathodic technique often yields hydroxides rather than oxides, making a post-deposition calcination for the transformation of the hydroxide to the oxide necessary. Unfortunately, the deposited amorphous hydroxides are often metastable. Other cathodic electrodeposition methods involve the direct reduction of the oxidation state of the metal cation and deposition of the oxide onto the electrode [9].

In contrast, the anodic oxidation technique relies on the metal cation exhibiting a low oxidation state that is anodically oxidized to a higher oxidation state [6]. Here the pH of the electrolyte is often critical and has to be chosen such that the lower oxidation state is stable while the higher oxidation state readily undergoes hydrolysis to yield the metal oxide or hydroxide. Another anodic method exploits the fact that the desired metal compound is precipitated on lowering the pH close to the electrode achieved by reduction of water or other additives.

In fact, some of the mentioned methods should be described as "electrochemically triggered chemical precipitation" rather than as "electrodeposition" because of the involved reaction mechanisms [6]. Unlike the electroplating of elementary metals, the deposition of metal oxides and hydroxides is usually not a simple Faradaic reaction, but a secondary precipitation induced by a primary product of an electrochemical reaction. Thus, when the concentration of metal cations at the electrode is not sufficient, the primary product, such as hydroxide ions, might diffuse away from the electrode surface resulting in deposit yields lower than the theoretical value predicted by Eq. (5.1) or in a porous and poorly connected deposit. In case of a templated electrodeposition, the sufficient supply of the working electrode with metal precursors might be affected by the mesoporous templates causing a non-dense deposit that is likely to collapse during template removal or calcination.

5.1.2 Electrochemical Intercalation

Intercalation is the reversible electromigration of mobile guest molecules, atoms or ions into a guest host or intercalation compound driven by an electric potential between the working and counter electrode [10]. The structural integrity of the crystalline host lattice that contains an interconnected system of empty lattice sites of appropriate size is only slightly perturbed and usually fully restored during guest deintercalation. The degree of solid state intercalation reactions or the concentration of ionic guest insertion, which are typically performed at low temperatures, is precisely controlled by monitoring the charge and the measured cell voltage, highlighting the kinetic control that is possible by electrochemistry [11, 12].

The interest in intercalation reactions stems from different motivations. From a preparation point of view they provide routes for the systematic synthesis of new solids with kinetic rather than thermodynamic stability that cannot be obtained by other preparation techniques [11, 13]. Furthermore, they permit controlled systemic modifications of chemical as well as physical properties, including electronic, magnetic and optical properties. From an application viewpoint, they are of importance in supercapacitors, rechargeable batteries, non-emissive electrochromic displays, and so forth [4, 14].

Remarkably, a large number of electronically conducting intercalation hosts are accessible today, including numerous transition-metal oxides with framework lattices, such as vanadium pentoxide and nickel oxide, and organic solids with molecular lattices. Additionally, a large variety of ionic guest species have been studied, including small ions like protons or metal ions, especially alkali ions, as well as large molecular ions. According to their essential lattice structures, intercalation compounds may be divided into three basic host lattice types with three-, two- and one-dimensionality, independent of their chemical composition [10]. In the case of three-dimensional framework structures, the size of the guest species which can be intercalated is constrained. In contrast, two-dimensional layered systems, characterized by a strong intralayer covalent bonding and relatively weak interlayer interactions which allow a fairly free widening of the interlayer separation, can accommodate guest species of larger sizes and higher concentration. Layered host lattices also allow for more structural flexibility providing the ability to adapt to the geometry of the guest species. In general, the incorporation of guest species is accompanied with structural changes which in case of layered systems cause modifications of the interlayer spacing, the stacking mode, and formation of intermediate phases exhibiting staging [15]. However, if intercalation is taken to its extreme, the layered host structure can completely delaminate, resulting in exfoliation and loss of the process reversibility [15].

The fundamental intercalation process can be defined as the bulk reaction of an electronically conducting host lattice (H) that contains an intracrystalline system of accessible vacancies ([ ]) with mobile guest ions (I) present in the electrolyte phase which is in contact with the solid driven by an applied electric potential. This transfer of ions between the electrolyte and the solid is accompanied by a compensating electron transfer [13]

$$[\ \ ]_x\,\mathrm{H} + x\mathrm{e}^- + x\mathrm{I}^+ \rightleftarrows [\mathrm{I}^+]_x\mathrm{H}^{x-} \tag{5.5}$$

$$[\ \ ]_x\,\mathrm{H} + x\mathrm{I}^- \rightleftarrows [\mathrm{I}^-]_x\mathrm{H}^{x+} + x\mathrm{e}^-. \tag{5.6}$$

This double injection of ions and electrons maintains electroneutrality of the system. The majority of intercalation compounds are known to undergo electron/cation transfer reactions, but NiO, for example, also supports reversible electron/anion transfer processes. In particular, an increase in the specific surface area is generally expected to significantly reduce the characteristic diffusion length of intercalation ions, while simultaneously increasing the number of accessible intercalation sites [16].

Orthorhombic crystalline vanadium pentoxide is a typical intercalation compound as a result of its layered structure, see Fig. 5.2, which finds widespread use in lithium ion intercalation applications such as electrochromic cells [17], high energy density batteries [18], supercapacitors [19], and sensors [20], since it offers the essential advantages of low cost, abundant availability, easy synthesis, and high intercalation densities [15, 16].

In 1975, Whittingham was first to report the reversible electrochemical lithium intercalation into vanadium pentoxide [11]. V_2O_5 electrodes are expected to undergo lithium ion insertion accompanied by several structural modifications, occurring in several consecutive stages [21, 22]

$$\mathrm{V_2O_5} + x\,\mathrm{Li}^+ + x\,e^- \leftrightarrow \mathrm{Li}_x\mathrm{V_2O_5}. \tag{5.7}$$

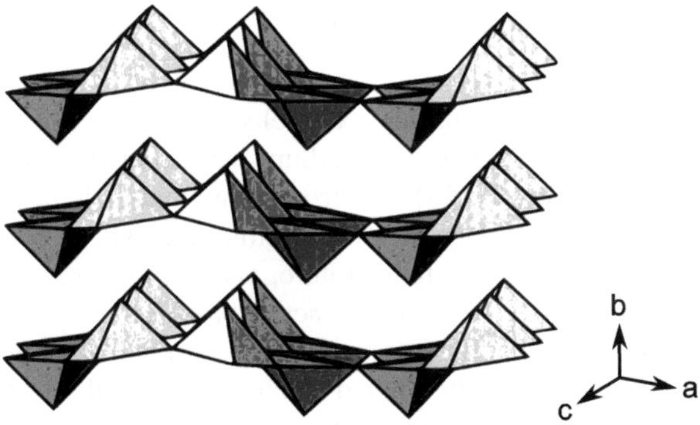

Fig. 5.2 Crystal structure of V_2O_5 consisting of layers of edge- and corner-sharing VO_5 square pyramids with the apical V–O bond distance being much shorter than the four other distances, corresponding to a double bond. (Reprinted with permission from Winter et al. [15]. Copyright 2013 WILEY-VCH Verlag GmbH)

The α-, ε-, and δ-phase exist for $x < 0.01$, $0.35 < x < 0.7$, and $x = 1$, respectively, and the transitions are fully reversible. The ε-phase is characterized by an increasing puckering of the layers, while δ-phase is formed by gliding of one layer out of the surrounding two [23]. Further lithium intercalation $x > 1$ results in an irreversible transformation to the γ-phase which itself can reversibly be cycled in the stoichiometric range $0 < x < 2$ without causing a structural change. Upon further lithiation $x = 3$, the irreversibly rock-salt ω-phase is formed. In the bulk, this bronze can be reversibly cycled as the ω-phase, but a large fraction of the lithium cannot be removed, causing a decrease in energy capacity [24].

The formation of a nanostructure in the intercalation compound not only enables the expansion and contraction of the host material during guest insertion and extraction, but also increases the surface to volume ratio. Recently, V_2O_5 synthesis focussed on the manufacture of nanostructured or mesoporous scaffolds with the aim to improve their lithium ion intercalation properties, which are limited by an inherently low diffusion coefficient ($D \leq 10^{-12}\,\mathrm{cm^2\,s^{-1}}$) and a moderate electrical conductivity ($\sigma = 10^{-2}$–$10^{-3}\,\mathrm{S\,cm^{-1}}$) [16].

Ion intercalation into nickel oxide is often performed in aqueous media. On immersion of nickel oxide into an alkaline solution, such as aqueous KOH, a spontaneous chemical conversion of NiO into a hydrous metal oxide phase $Ni(OH)_2$ proceeds [25]. Hydrous NiO resembles V_2O_5 in respect to the two-dimensional layered host structure. The Ni(II)/Ni(III) couple responsible for the reversible color change on electrochemical cycling has been identified as the nickel hydroxide/oxy-hydroxide phases [25, 26]

$$\mathrm{Ni(OH)_2} \rightleftarrows \mathrm{NiOOH} + \mathrm{H^+} + \mathrm{e^-}. \tag{5.8}$$

5.1.3 Metal Oxide Electrochromism

Following Deb's discovery of tungsten oxide electrochromism in 1969, numerous other transition-metal oxides, such as vanadium pentoxide and nickel oxide, were found to be electrochromic [27, 28]. Electrochromic materials belong the class of chromogenic materials that change their absorption spectra as response to chemical or physical stimuli, such as light, temperature, and pH. The reversible, persistent electrochromic change in color or transparency is based on the change of the metal oxide's oxidation state or electron densities during ion intercalation/deintercalation and is controlled by a temporarily applied electrical potential. For this purpose an operation potential of only a few volts is sufficient to control the direction and magnitude of the transferred charge and the resulting optical properties. Moreover, little or no input of power is needed to maintain a specific redox state, which leads to the desired optical memory effect and low energy consumption [29, 30]. Additionally, because of their simple multilayer design, large surface area electrochromic devices can be cheaply manufactured [30]. Typically, the conducting metal oxide film forms the chromogenic electrode, which is separated from the charge balancing counter electrode by an electrolyte. As low-cost, low-power, and low-voltage technology it is highly attractive for a wide range of potential applications.

While electrochromic MO has been employed in smart windows and display devices, only very few electrochromic products are commercially available [31], despite extensive research for over forty years [27]. This is mainly due to the fact that the employed conventional flat film devices suffer from limited ion intercalation and therefore from a poor electrochromic performance such as long switching times, low chromatic contrasts, and unsatisfying coloration efficiencies [13]. The chromatic contrast $\Delta T(\lambda) = T_b - T_c$ between the colored and bleached state is limited by finite ion penetration depths, while the switching time τ is restricted by the long diffusion distances required to achieve a satisfactory color saturation [32]. As a consequence the conventional devices exhibit an unsatisfyingly low composite coloration efficiency $CCE = \Delta OD/q$, which is defined by the charge per area q that was required to induce a gradient in optical density $\Delta OD = \ln(T/T_0)$, where T_0 is the initial transmittance. The ideal electrochromic material would display a large transmittance change with a small amount of charge, giving rise to a high CCE [28, 30, 33]. A simple solution to improve the electrochromic performance is to enhance the intercalation properties by the formation of a network structure [34].

Nanomaterials of undoped vanadia employed in electrochromic studies include nanowires [35–37], inverse opals [34], and mesoporous films [38]. Similar attempts were undertaken for NiO devices [39–42]. Although these recent attempts brought about an improvement in electrochromic performance, the reported switching times remained 1–2 orders of magnitude above the desired video rate of 24 frames per second [34–38]. This is mainly due to the use of sub-optimal MO morphologies in terms of their structural dimension, connectivity and integrity.

5.1.4 Metal Oxide Supercapacitors

Supercapacitor charge storage offers a number of desirable properties when compared with conventional batteries, including charging within seconds, long-term cycling stability and the ability to deliver up to ten times more power. These features are used not only in high power demanding applications such as power buffers and power saving units, but also in energy recovery. Whereas charge storage through non-faradaic means is fairly straightforward, a variety of distinct chemical options are available for charge storage through faradaic or redox-type reactions.

There are two types of supercapacitors, differentiated by their charge storage mechanism. In the electrochemical double layer capacitor (EDLC), charge is stored electrostatically by the reversible and very rapid adsorption and desorption of ions on/from high-surface-area materials, for example, activated carbon. The charges accumulate in the electrical double layer at the electrode/electrolyte interface. A major limitation of traditional EDLCs is their low energy density, which accounts for the increased interest in the use of redox materials for electrochemical capacitors with energy densities that are higher by more than one order of magnitude. In the second type of supercapacitor, known as a pseudocapacitor, kinetically facile faradaic reactions occur at the electrode surface and electrons are transferred to or from the valence bands of the redox cathode or anode reagent [43]. Capacitive contributions arise from charge-transfer processes with surface atoms as well as near-surface ion intercalation, referred to as redox and intercalation pseudocapacitance, respectively. The definition of pseudocapacitance requires that both these faradaic processes are kinetically facile and not diffusion controlled on the time scale of interest [44]. Such surface faradaic reactions are slower than ionic adsorption/desorption, resulting in slower charge/discharge rates compared to EDLCs, but faster than in batteries, where faradaic reactions take place in the bulk of the electrode, rather than at the electrode surface [45]. However, diffusion controlled ion insertion and double layer formation might also contribute to the total charge stored in pseudocapacitors [46].

The specific capacitance C, energy density E, and corresponding power density P are defined by the following equations [4]

$$C = \frac{I}{m}\left(\frac{\mathrm{d}V}{\mathrm{d}t}\right)^{-1} \tag{5.9}$$

$$E = \frac{1}{2}CV^2 \tag{5.10}$$

$$P = \frac{V^2}{4Rm}, \tag{5.11}$$

where I is the constant discharging current, m is the mass of the total active electrode material, V is the cell potential, R is the equivalent series resistance, and $\frac{\mathrm{d}V}{\mathrm{d}t}$ is the slope of charge-discharge curve typically obtained from galvanostatic measurements.

Growing interest in electric vehicles allied with the ubiquity of increasingly capable mobile devices create a seemingly insatiable demand for electrochemical energy storage with higher power and energy densities. These two requirements, unfortunately, cannot be met by current energy storage devices. While lithium batteries provide high specific energy densities typically on the order of 100–1,000 $Wh\,kg^{-1}$, they deliver only limited specific power upon discharge on the order of 0.1 $kW\,kg^{-1}$. Supercapacitors have the ability to deliver significantly more power but currently store less energy. The typical specific energy and power densities of EDLCs are on the order of 1 $Wh\,kg^{-1}$ and 1 $kW\,kg^{-1}$, respectively. These specific values are not high enough to replace batteries in many applications. Supercapacitors based on nanostructured transition metal oxide architectures have been developed to allow both redox and intercalation pseudocapacitance. This approach has the potential to simultaneously optimize the energy and power densities in a single material. One way to increase pseudocapacitance is to improve the electrochemical intercalation properties by an increase of the electrode surface area. As the structure of electrochemically active materials approaches nanoscale dimensions, pseudocapacitive effects become increasingly important. The dependence of pseudocapacitance on the size of anatase TiO_2 nanocrystals has been established quantitatively [46]. The results showed that for particles smaller than 10 nm in diameter, more than half of the total charge is stored via pseudocapacitive processes.

Because of the various advantages mentioned previously, vanadia is also highly important as supercapacitor electrode material. V_2O_5 aerogels can act as high rate, high capacity hosts for lithium, because they have a very high surface area and the diffusion distance that must be penetrated by lithium is very short [47]. Amorphous V_2O_5 in a nanoporous alumina template showed a high specific capacitance of 900 $F\,g^{-1}$ in 1 M KCl electrolyte [48]. Promising capacitive behavior was observed for hydrous vanadium oxide with specific capacitance reaching 737 $F\,g^{-1}$ in 12 M LiCl for a potential window between -0.2 and 0.8 V [34]. Electrospun V_2O_5 nanofibers as supercapacitor electrodes showed a promising energy density of 5 $Wh\,kg^{-1}$ [49]. V_2O_5 has also been applied in lithium batteries, with a discharge-charge efficiency up to 97.2 % in the potential range from 2 to 4 V [50]. High performance supercapacitors based on intertwined carbon nanotube/V_2O_5 nanowire composites exhibit specific energy capacity of 40 $Wh\,kg^{-1}$ and a power density of 210 $W\,kg^{-1}$ [51]. Binder-free flexible supercapacitors containing vanadium oxide nanowires and carbon nanotubes showed further improvement in power density (5.26 $kW\,kg^{-1}$) and energy density (46.3 $Wh\,kg^{-1}$) [52]. High intercalation rates of V_2O_5 gel/carbon composites in supercapacitors were also reported with a specific energy density of 20 $Wh\,kg^{-1}$ [19]. Pseudocapacitors made from V_2O_5-coated self-standing carbon-nano-fiber paper achieved a specific capacitance per mass of vanadia of 1, 308 $F\,g^{-1}$ in a 2 M KCl electrolyte, though this high value decreases to 214 $F\,g^{-1}$ when considering the total mass of the composite electrode [53]. Three-dimensional bicontinuous ultrafast-charge and discharge bulk battery electrodes made by electrodeposition through colloidal templates have been reported [54].

Ideal supercapacitor behavior with large specific capacitance ($\approx$350 $F\,kg^{-1}$) of amorphous $V_2O_5 \cdot nH2O$ in aqueous solution containing KCl with optimized pH has been reported [55]. However, the application of water in supercapacitors limits their operation temperature as well as the electrochemical window. Using room temperature ionic liquids (RTILs), molten salts with a melting point close to or below room temperature, avoids these limitations. RTILs are composed of bulky organic cations and smaller anions that only loosely associate. Their good solvation properties, high conductivity, non-volatility, low toxicity, and good electrochemical stability make RTILs suitable for many electrochemical applications. Non-flammable RTIL gels have been used as electrolyte in supercapacitors, enabling flexibility, safe operation, and a large electrochemical window (i.e. the electrochemical potential range over which the electrolyte is neither reduced nor oxidized at the electrodes) [56]. Unlike organic solvents, the high boiling point of RTILs makes the encapsulation of the device easier and avoids electrolyte evaporation during testing, thereby further enhancing the cyclability of the device.

5.1.5 Concluding Remarks

It was stressed above, that for certain applications a conceptually ideal intercalation compound should be a mesoporous fully interconnected network consisting of struts with a homogeneous diameter and a narrow pore diameter distribution, enabling effective electrolyte infiltration and good charge transport. Small pore diameters (2–50 nm) give rise to dense highly porous films which can be made thin enough so that the assembled device has a low electrical resistivity. To enable fast electrochemical and therefore electrochromic response times below 100 ms, the strut radius in the nanostructured V_2O_5 device should be comparable to the lithium diffusion distance $x = \sqrt{2Dt} \approx 5$ nm. Further reduction of the structural feature size is probably counterproductive because of the lowered conductivity, arising quantum effects, and lack of structural stability of the network. The double-gyroid morphology is a promising candidate to fulfil the listed requirements [57].

5.2 Experimental Methods

5.2.1 Materials

The chemicals and materials listed in Table 5.1 were used as received unless stated otherwise.

5.2.2 V_2O_5 Electrodeposition and Thermal Annealing

The electrolyte used for the electrodeposition [37, 58] was a 1 M aqueous vanadium sulfate oxide hydrate solution containing 50 $\%_{\mathrm{wt}}$ ethanol with a pH of 1.6. A standard

Table 5.1 Chemicals and materials used

Chemical	Purity (%)[a]	Abbreviation
Bis(trifluoromethane)sulfonimide lithium	– (SA)	LiTFSI
Ethanol, laboratory reagent grade	96 (SA)	–
Ethylenediaminetetraacetic acid	98 (SA)	EDTA
Lithium	99 (SA)	–
Lithium perchlorate	95 (SA)	–
Nickel chloride	98 (SA)	–
Propylene carbonate, anhydrous	99.7 (SA)	–
Silver, wire 0.25 mm diameter	99.9 (SA)	–
Triethylsulfonium bis (trifluoromethylsulfonyl) imide	99 (SA)	[SET3][TFSI]
Vanadium sulfate oxide hydrate	99.9 (AA)	–
Water, deionized 18 MΩ	–	DI water
Zinc nitrate hexahydrate	98 (SA)	–

[a] Suppliers are given in brackets: Alfa Aesar, AA; Sigma Aldrich, SA

three-electrode electrochemical cell with a Pt mesh or stainless steel counter electrode and Ag/AgCl reference electrode was employed. The FTO substrates to be coated were immersed into the electrolyte a couple of minutes before starting the electrodeposition process until a stable open circuit potential was reached, in case of template-coated FTO this indicated complete electrolyte infiltration into the templates. The low surface tension of ethanol thereby enabled the infiltration of the hydrophobic PXS pores. Anodic deposition of V_2O_5 was performed at a constant potential of 1.5 V using an Autolab PGSTAT302N potentiostat. The overall reaction can be summarized as [58]

$$2VO^{2+} + 3H_2O \rightarrow 4V_2O_5 + 6H^+ + 2e^-. \tag{5.12}$$

The gradually increase in film thickness was controlled by monitoring the deposition time rather than the charge passing through the cell because of the reason mentioned in Sect. 5.1.1.2. Subsequently, the vanadium oxide films were rinsed with deionized water, dried, and in case of templated deposition were freed of the styrenic scaffold by dissolution in toluene.

The samples were then heated to temperatures between 200 and 400 °C at a ramp rate of 1 °C min^{-1} on a hotplate and annealed for two hours in air. An annealing temperature of 275 °C was found to be optimal for the preparation of electrochromic and supercapacitor electrodes, as will be discussed in Sect. 5.3.1.2 [59, 60].

5.2.3 Electrochromic Device and Supercapacitor Assembly

Transparent electrochromic cells consisting of a FTO/V_2O_5/$LiClO_4$+PC/FTO layer sequence and an Ag/AgCl wire as reference electrode were assembled by

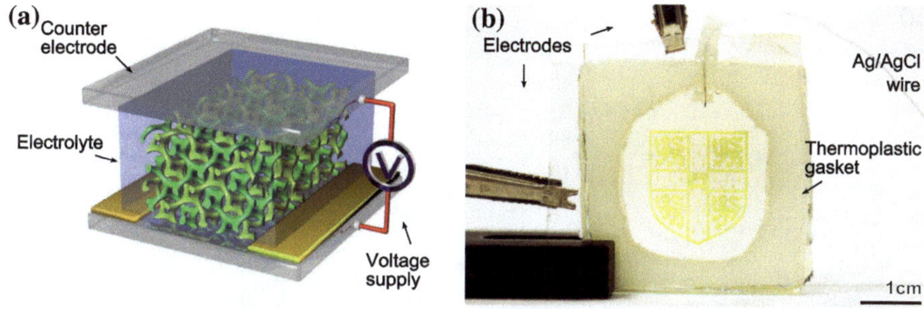

Fig. 5.3 Electrochromic cell assembly. **a** Schematic illustration of the layered device design based on a DG-structured electrode. **b** Functional electrochromic nano-device with the design pattern in the chromogenic double-gyroid film showing the crest of the University of Cambridge. The thickness of the vanadia deposit was measured to be around 1.1 μm

capping a V_2O_5-coated electrode with a plain, piranha cleaned FTO counter electrode using a precut thermoplastic gasket (Parafilm) as spacer, see Fig. 5.3. The cells were fused by melting the thermoplastic gasket at 150 °C for 30 s after insertion of an Ag/AgCl wire which later served as reference electrode. After assembly, the electrolyte, a 1 M $LiClO_4$ propylene carbonate solution, was infiltrated using a syringe and the cell was sealed with epoxy glue. The Ag/AgCl wire was prepared by electrochemically oxidizing a silver wire in a 3 M potassium chloride solution at a current density of 10 mA cm^{-2} for 15 s.

Similarly, supercapacitor cells were fabricated by sandwiching a precut gasket between a pair of laterally offset V_2O_5 electrodes, obviating the need for a separator and enabling the observation of color change of both electrodes. Instead of propylene carbonate the highly conducting ($\approx$5.5 mS cm^{-1}) room temperature ionic liquid [SET3][TFSI] in combination with the corresponding lithium salt, LiTFSI, was chosen as electrolyte and the integration of Ag/AgCl reference electrode was omitted. Figure 5.4 shows the supercapacitor design consisting of FTO/V_2O_5/LiTFSI+[SET3][TFSI]/V_2O_5/FTO layer sequence. Further, a battery-driven stand-alone device, see Fig. 5.18, was assembled.

5.3 Results and Discussion

5.3.1 Preparation of Metal Oxide Electrodes by Electrodeposition

5.3.1.1 Zinc Oxide and Nickel Oxide

Three-dimensionally nanostructured transparent conducting ZnO could serve as electrode for several applications such as dye-sensitized solar cells. Thus, the template-assisted nanopatterning of ZnO via electrodeposition was attempted, following the

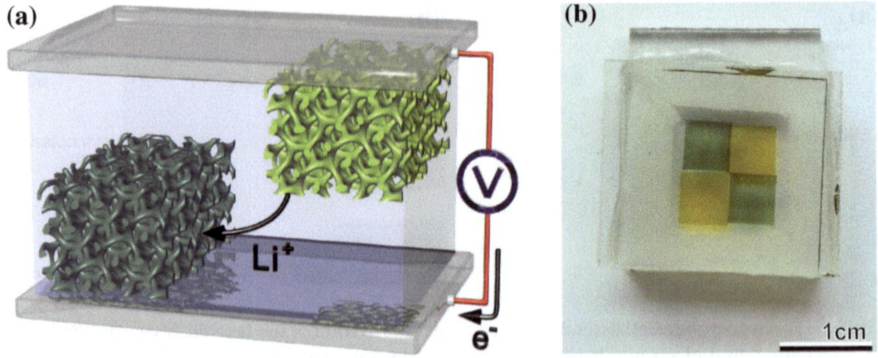

Fig. 5.4 Supercapacitor cell assembly. **a** Schematic illustration of the supercapacitor design based on two laterally offset double-gyroid structured electrodes. **b** Photograph of a fully transparent electrochromic supercapacitor consisting of an oxidized yellow top electrode, a laterally offset green/gray bottom electrode in the reduced state

recipe of Han and Tao for the deposition of highly transparent and conducting yttrium-doped ZnO using 0.1 M zinc nitrate electrolyte with varying concentrations of yttrium nitrate [61]. The reported, nontemplated deposition on bare FTO was reproducibly yielding micron-sized ZnO crystals. However, the templated deposition using DG scaffolds was unsuccessful because the dominant crystal formation at the interface between the FTO substrate and template led to the delamination of the latter.

Next, the template-assisted nanostructured deposition of NiO was tried, following the experimental method of Sonavane et al. for the electrodeposition of nontemplated coatings [62]. For this purpose a 0.5 M aqueous NiCl bath containing 0.1 M KCl was prepared, which was complexed using EDTA and pH-adjusted to 8 by addition of KOH. Electrodeposition was conducted in a three-electrode cell at a potential of -1.1 V versus Ag/AgCl for 100 min. The obtained DG-structure NiO film of about 1 μm thickness is presented in Fig. 5.5. This preparation technique has two major disadvantages: firstly, the very slow deposition rate, and secondly, the transparency of the deposit complicates the anyway difficult preparation route. In Chap. 6, a more elegant approach for nanostructured NiO deposition is presented that overcomes these issues.

5.3.1.2 Vanadium Pentoxide

Initially, electrolytes containing no or only a small fraction of ethanol and less than 1 M $VOSO_4$ were tested. While these electrolytes gave homogeneous coatings on plain FTO, unsatisfying results were obtained when refilling voided DG templates. The film thickness of the templated coating varied locally resulting in a rough free surface, see Fig. 5.6a. Further, the deposit was not dense and therefore showed poor stability. These issues could be resolved by optimizing the vanadia precursor

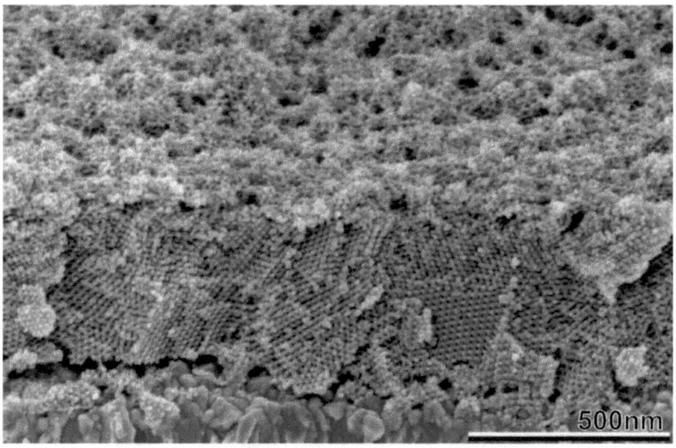

Fig. 5.5 DG-structure NiO film prepared by templated electrodeposition and subsequent template removal by dissolution in toluene

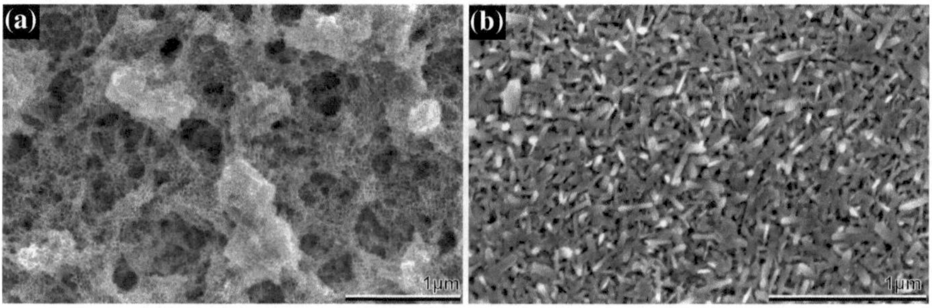

Fig. 5.6 **a** SEM of the free-surface of a templated V_2O_5 film of poor quality which was deposited using an electrolyte with a low ethanol and $VOSO_4$ concentration. **b** Needle-like V_2O_5 crystallites formed during thermal annealing of a DG-structured vanadia film at 400 °C destroying the initially existing nanostructure

concentration and the water-ethanol ratio Sect. 5.2.2 [38]. A suitable potential for the potentiostatic V_2O_5 deposition was determined by analyzing cyclic voltammograms (CVs) using bare and templated-coated FTO substrates as working electrodes, see Fig. 5.7. In case of the templated-coated FTO electrode, the current reaches a plateau at $\approx$1 V, thus a deposition potential of 1.5 V was identified as suitable.

The open circuit potential of fully porous double-gyroid templates on FTO when immersed in electrolyte was 480 $\pm$ 10 mV. A deposition rate of approximately 9.2 nm s^{-1} was observed for the replication of gyroid templates. The deposited film thickness was measured by SEM imaging of prepared cross-sections. The typical current variation during deposition is shown in Fig. 5.7c. Similarly, nontemplated and inverse opals (IO) were prepared under the same conditions, except that the required deposition time for the same deposition thickness was much shorter. The deposition rate in colloidal arrays made from poly(styrene) microspheres with a diameter of 400 nm, was approximately 60 nm s^{-1}.

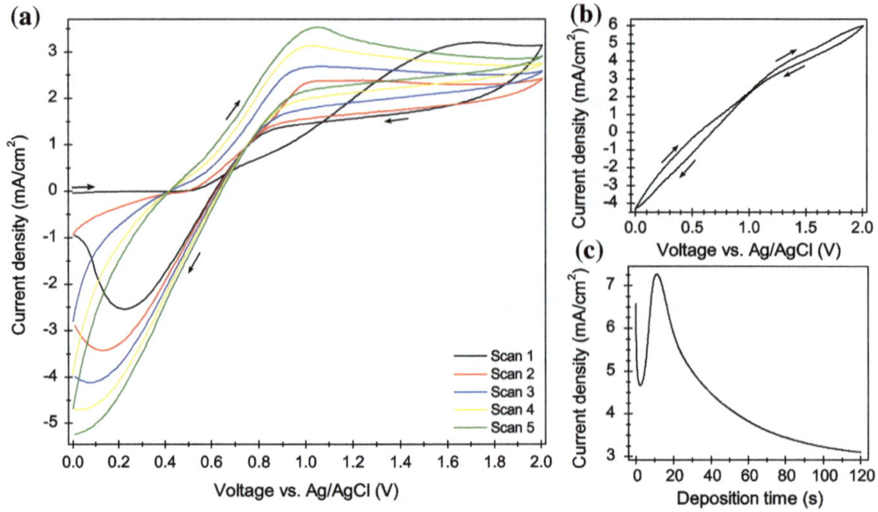

Fig. 5.7 Electrochemical deposition of V_2O_5 from 1 M $VOSO_4$ electrolyte. **a** CV curves for a DG template on a FTO substrate recorded at a scan rate of 50 mV s^{-1}. The current density of the DG template reaches a plateau at ≈1 V which is most likely caused by diffusion limitations within the confining DG channels. **b** Corresponding CV for a bare piranha cleaned and silanized FTO electrode. In contrast to **a** no current density plateau is observed. **c**, Variation of the current density during refilling of a DG template with V_2O_5 with a constant potential of 1.5 V applied. A total charge of 508.8 mA cm^{-2} was passed through the cell during deposition to yield a 1.1 ± 0.1 μm thick nanostructured deposit

Using SU-8 patterned FTO substrates to create visible design patterns and small area electrodes worked exceptionally well, see Fig. 5.3 [63]. Additionally, limiting the deposition area with a SU-8 mask allowed an improved control over the templated electrodeposition yielding defect-free deposits of homogeneous thickness, see Fig. 5.12.

Interestingly, voided films prepared from polymers such as PS50-*b*-PLA214$|^{173\,°C}_{0.33\,h}$ which formed a layer of lying cylinders at the free-surface, see Fig. 4.15c, offered the possibility to fabricate nanostructured vanadia coatings with very smooth free-surfaces. When using those films as templates the V_2O_5 electrodeposition was only aborted after a current increase was observed, indicating the onset of overgrowth after complete refilling of the template. Since the free-surface of these templates only exhibits a few pores, the overgrowth tended to be loosely attached and could be rinsed away using DI water. The free-surface of vanadia deposits prepared in this fashion nicely revealed which double-gyroid planes are formed underneath the layer of lying cylinders, see Fig. 5.8d. The smooth free-surface might be of interest for optical applications, since rough surfaces tend to increase light scattering.

Directly after deposition the films were rinsed and immersed into DI water for several minutes. Failing to do so or using ethanol instead of water resulted in nonporous V_2O_5 free-surfaces as depicted in Fig. 5.8. In order to improve the crystallinity of

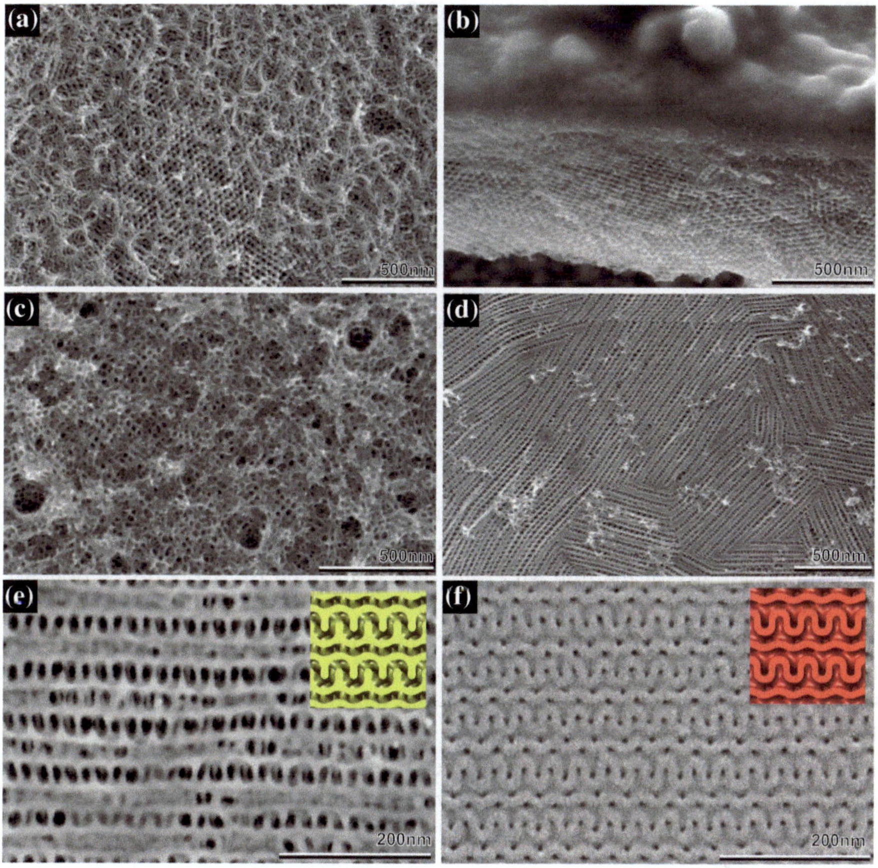

Fig. 5.8 SEM images of the DG-structured V_2O_5 electrodes after annealing at 275 °C. **a** Highly porous film/substrate interface showing the topography of the FTO substrate revealing the existence of a worm-like region between the FTO substrate and the ordered DG-structured bulk. **b** Failing to rinse the samples after V_2O_5 electrodeposition led to a nonporous free-surface, evident in this cross-section. **c** Typical free-surface of a partially refilled DG template. **d** Vanadia free-surface obtained by fully replicating a PS50-*b*-PLA214$|_{0.33\,h}^{173\,°C}$ template, showing the inverse double wave pattern of the (211) plane. **e** Magnified view of **d**. **f** For comparison, the free-surface of a voided P(F)S49-*b*-PLA192$|_{0.33\,h}^{173\,°C}$ is shown, which also shows the double wave pattern (reprinted from Fig. 4.15). Simulations of the (211) plane for the majority and minority DG phase with an unit cell dimension of 42 nm are displayed as insets

the hydrous V_2O_5 and to remove the intercalated water, the films were thermally annealed at temperatures between 200 and 400 °C. The DG structural integrity persisted up to about 300 °C. Annealing temperatures above this value led to the formation of crystallites larger than the gyroid strut size and therefore the destruction of the nanomorphology, see Fig. 5.6b. Based on these results, 275 °C was chosen as annealing temperature for the preparation of nanostructured V_2O_5 electrodes.

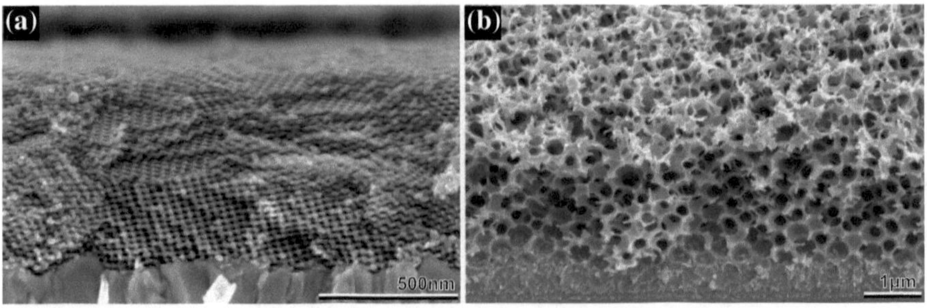

Fig. 5.9 Cross-sectional SEM images of templated V_2O_5 electrodes highlighting the differences in structural dimensions. **a** Mesoporous DG-structured deposit. **b** Macroporous IO structure. *Note* the different scale bars

The highly periodic DG V_2O_5 replica formed upon removal of the styrenic template by dissolution was fully preserved during this annealing step, see Figs. 5.8 and 5.9.

The green color of the as-deposited vanadium oxide was maintained during temperature annealing at 275 °C and indicated the presence of V_2O_5. The stoichiometry and crystallinity of the thin MO films was confirmed by X-ray diffraction, see Fig. 5.10.

Given that the density of electrodeposited nontemplated vanadia is $\approx 2.87\ \mathrm{g\ cm^{-3}}$, the average density of double-gyroid structured vanadia with a volume fraction of 37.9 % is $\approx 1.09\ \mathrm{g\ cm^{-3}}$, illustrating the porous nature of the DG structure [16]. Thus, the specific surface area of DG-structured V_2O_5 is $\approx 1.48\ \mathrm{m^2\ g^{-1}}$. Furthermore, the mass m of vanadia electrodes was calculated using these densities, the electrode area and thickness. The V_2O_5 film thickness was determined by cross-sectional SEM or with a surface profilometer.

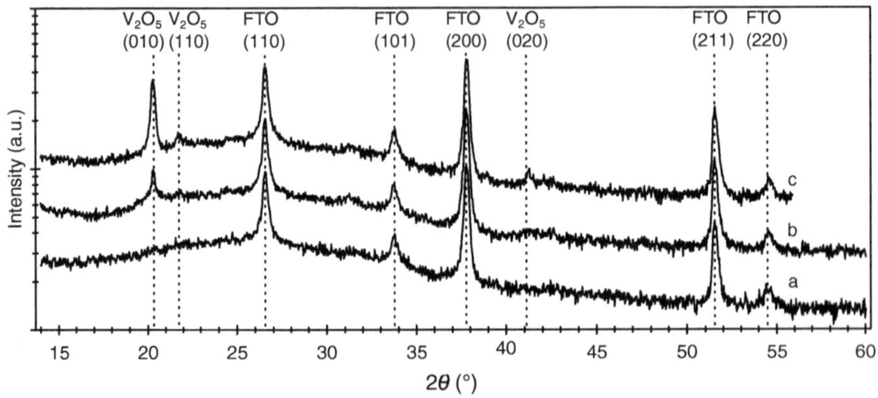

Fig. 5.10 X-ray diffraction patterns of a bare FTO substrate (**a**), a DG-structured (**b**) and non-templated (**c**) vanadium pentoxide film annealed at 275 °C for 2 h. Cu K_α radiation was used and the patterns were matched with the corresponding tetragonal SnO_2 (PDF number: 046-1088) and orthorhombic V_2O_5 (PDF number: 089-0611)

5.3.2 Electrochemical Characterization of V_2O_5 Electrodes

The electrochemical Li insertion/extraction behavior of the V_2O_5 electrodes in 1 M $LiClO_4$ propylene carbonate solution was investigated by cyclic voltammetry using a three-electrode electrochemical cell. The V_2O_5 samples formed the working electrode and three types of V_2O_5 structures were studied: a DG-structured, an IO-structured, and an unstructured compact layer electrode. The experiments were conducted with two different pairs of counter and reference electrodes. Firstly, the CVs for a DG and an IO electrode were recorded using a FTO substrate as counter electrode and an Ag/AgCl reference electrode, see Fig. 5.11a. Secondly, lithium foils were used as both the counter and reference electrodes under a protective argon atmosphere to measure the CVs for a DG and a nontemplated V_2O_5 deposit, displayed in Fig. 5.11b. The current density discrepancy between the two measurements arises from different scan rate used. All CVs show the two well-resolved sets of redox peaks occurring during lithium ion intercalation and extraction according to Eq. (5.7) [17]. The redox peaks of the double-gyroid electrodes are higher and sharper than those of the IO and nontemplated ones. This indicates enhanced lithium ion intercalation and consequently, improved electrochemical properties for devices based on the nanostructured DG electrodes.

In the following, the CVs of Fig. 5.11B are discussed in more detail. For the DG-structured V_2O_5 electrodes (Fig. 5.11Bb), only a fraction of the V^{5+} ions are reduced to V^{4+} ions in the first reduction peak at about 3.4 V. This reduction changes the color of the material from yellow to green. The remaining V^{5+} ions are reduced to V^{4+} in the second peak at 3.2 V. The reverse reactions give rise to the two corresponding oxidation peaks in the reverse scan. Two phases of $Li_xV_2O_5$ are typically formed in this electrochemical reaction. The binary phase diagram of V_2O_5 - $Li_xV_2O_5$

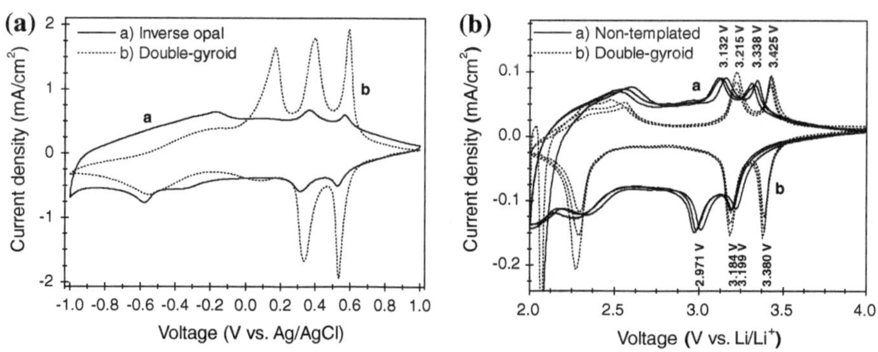

Fig. 5.11 Cyclic voltammograms V_2O_5 electrode in 1 M $LiClO_4$ propylene carbonate solution. **a** CVs of IO (*a*) and DG (*b*) electrodes recorded at a scan rate of 20 mV s^{-1} using a FTO counter electrode and an Ag/AgCl reference electrode. Kim et al. attribute the additional peaks between 0 and 0.25 V to lower oxidation states of vanadium (V^{4+}) [64]. **b** CVs of nontemplated layer (*a*) and DG (*b*) electrodes recorded at a scan rate of 1 mV s^{-1} using lithium foils as both the counter and reference electrodes

indicates that the peaks at 3.2 V and 3.4 V correspond to the phase transformation from α-V_2O_5 to ε-V_2O_5 and from ε-V_2O_5 to δ-V_2O_5, respectively [23]. For the DG-structured V_2O_5 the color changes from green/gray to yellow when the voltage is above 3.5 V.

The two sample types differ in the location and separation distance of the CV peaks. For the same experimental conditions, the anodic voltage peaks lie at $\approx$3.4 V and $\approx$3.2 V for DG-structured V_2O_5, more than 200 mV higher than the nontemplated layer with $\approx$3.2 V and $\approx$3.0 V, respectively. The average voltage difference between the cathodic and the corresponding anodic double-peak of the redox reactions in Fig. 5.11b decreases from 150 mV for the nontemplated layer to 38 mV for the DG material. A further reversible redox process is observed in the 2–2.8 V potential range, where a second lithiation step takes place. This deep lithiation occurring at 2 V changes the color from green to blue and may cause structural change. These structural changes have no detrimental effect on the reversible lithium insertion at capacities corresponding to the uptake of 2 Li atoms per V_2O_5 in the 2–4 V potential range [22].

5.3.3 Electrochromic Displays Based on V_2O_5 Electrodes

To investigate the role of the structure size, electrochromic devices were prepared from templated V_2O_5 electrodes with double-gyroid as well as inverse opal (IO) morphologies [34]. The IO electrodes were prepared from colloidal arrays of microspheres with a diameter of 400 nm. The electrochromic color change was quantified by time-resolved optical measurements on V_2O_5 electrodes with a small active area of 0.95 mm^2. Importantly for the determination of the figures of merit, these electrodes were characterized by a homogenous thickness and a well-defined area which was accurately measurable using a calibrated microscope. However, compared to large area displays, these cells showed enhanced switching kinetics.

Cycling the DG device from Fig. 5.3b between -1 V and $+1$ V gave rise to a color change from blue-gray to green-yellow as shown in Fig. 5.12c. Transmission spectra

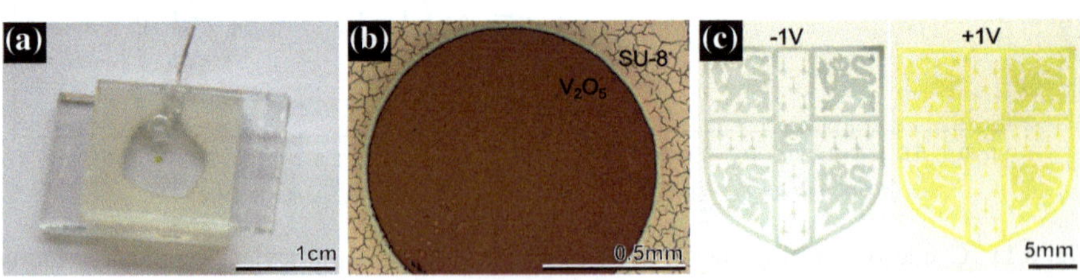

Fig. 5.12 **a** Photograph of a typical cell used for the electrochromic test with a well-defined round V_2O_5 area of 0.95 mm^2. **b** Micrograph of the electrochromic spot of (**c**) showing a homogenous appearance. **c** Photographs of electrochromic area in the *bluish gray* reduced (-1 V) and the *yellow-green* oxidized state (1 V) of the device displayed in Fig. 5.3b

were recorded with a spectrometer attached to a microscope using an optical fiber with a core diameter of 50 μm. The integration time was set to 25 ms. Figure 5.13 and Table. 5.2 summarize the obtained data for DG-structured electrodes of 0.73 μm and 1.1 μm thickness, termed $^{0.73\,\mu m}$ DG and $^{1.1\,\mu m}$ DG respectively, and compares their performance to $^{1.2\,\mu m}$ IO and $^{1.8\,\mu m}$ IO electrodes, correspondingly. Thereby, DG are juxtaposed with IO that show approximately the same transmission values at $\lambda = 430$ nm for the colored oxidized state.

Figure 5.13a shows the optical transmittance spectra after application of a ±1 V potential to the devices for 2 s. The cathodic potential (−1 V) causes lithium intercalation into V_2O_5 according to Eq. (5.7) and bleaching of the electrode. In the DG-structured samples, this gives rise to a fairly homogeneous transmission of ~50 % across the entire spectral range. Charge extraction under an anodic potential of +1 V causes a very high transmission above 500 nm with substantial absorption in the 400–500 nm range, causing the yellow-green coloration. While the color contrast of the DG devices benefits from a pronounced variation across the entire wavelength spectrum, the color change in the IO structure stems only from the 400–500 nm range. For visible wavelength, the $^{1.1\,\mu m}$ DG device shows a maximal transmission variation of $\Delta T = 49.8\,\%$ at $\lambda = 430$ nm where the $^{1.8\,\mu m}$ IO only reaches a value of $\Delta T = 14.1\,\%$. Slightly higher values were obtained for the other two samples, see Fig. 5.13C. The transmittance at $\lambda = 430$ nm is about 10 % higher for the thin electrodes compared to the one-third thicker V_2O_5 deposits. Curiously, for the colored state the $^{0.73\,\mu m}$DG electrode exhibits a lower transmittance in the wavelength range from 600–900 nm than the thicker $^{1.1\,\mu m}$DG sample.

A second important aspect of electrochromism is the temporal response under alternating 2 s potential steps (±1 V). The devices based on DG electrodes showed sharp and distinct transitions between the colored/oxidized and bleached/reduced state, see Fig. 5.13b. This switching behavior was analyzed in more detail by monitoring the transmittance at 430 nm (Fig. 5.13c). The characteristic response time τ was determined by fitting the exponential function $T(t) = T_0 + \Delta T \cdot e^{-t/\tau}$ to the switching curves. The DG devices showed short switching times below 82.5 ± 2.9 ms for the bleaching step and 86.7 ± 2.3 ms for the reverse process. In contrast, the IOs showed a much slower temporal response with switching times above 88 and 205 ms for a considerably smaller change in transmittance.

A further important requirement of electrochromic materials is cycle stability and stability of the structural integrity of the nanostructured material. Both investigated structures exhibited good long-term cycle stability with less than 20 % drop in coloration contrast after hundreds of switching cycles. The V_2O_5 nanostructure remained intact during extended cycling, as confirmed by SEM.

Finally, the charge per area q that was required to induce a gradient in optical density $\Delta OD = \ln(T/T_0)$ was measured, where T_0 is the initial transmittance. In (Fig. 5.13d) the optical density $|\Delta OD|$ at 430 nm is plotted against charge density $|q|$ during ion intercalation (orange) and extraction (blue). Significantly higher changes in optical density were observed for the gyroid device compared to the inverse opal for equal amounts of charge density passed through the cells. As discussed in Sect. 5.1.3 the ratio of these two quantities defines the composite coloration efficiency CCE.

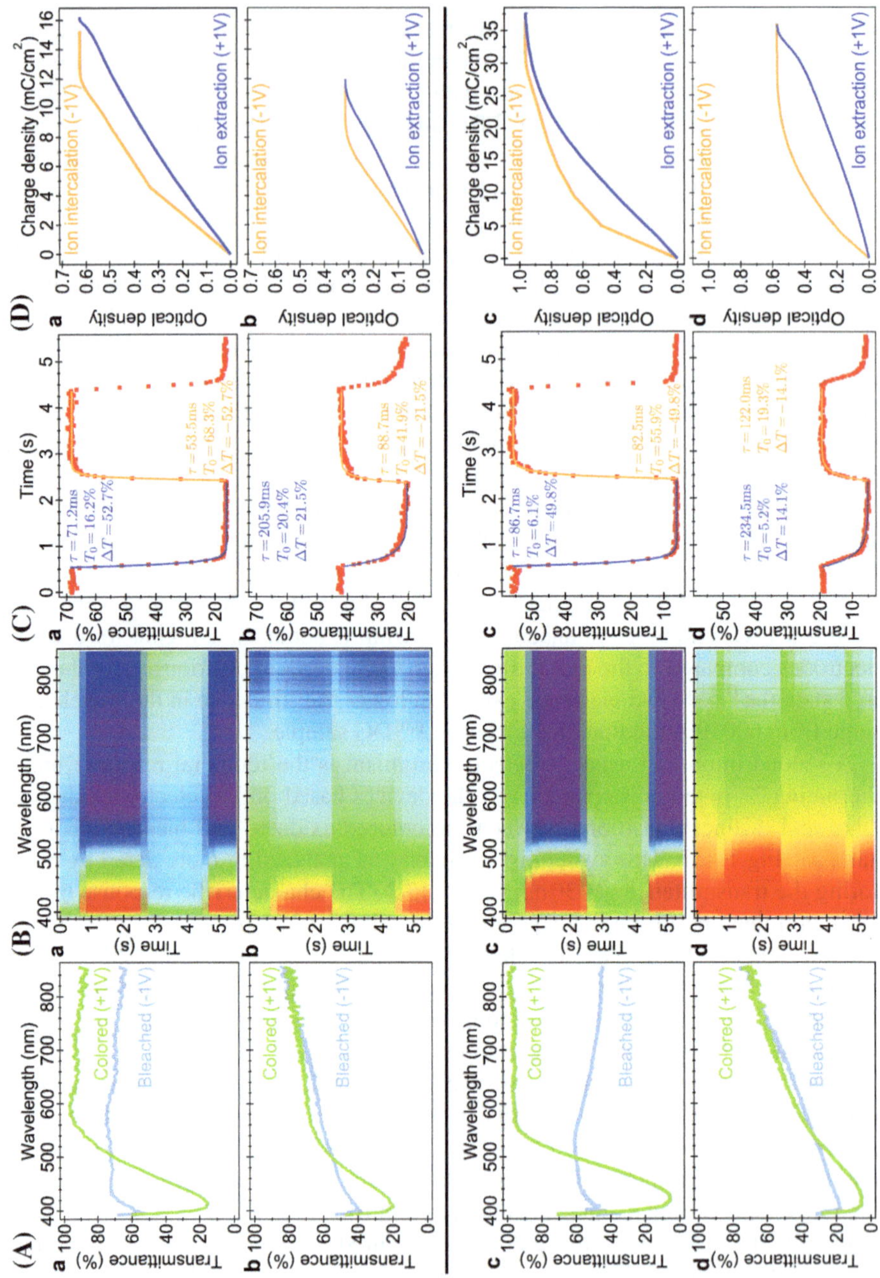

Fig. 5.13 **a** Optical transmittance spectra. **b** Time-dependent transmittance variation for the visible spectrum. **c** Switching curves at $\lambda = 430$ nm. **d** Optical density $|\Delta OD|$ at 430 nm versus charge density. For $^{0.73\,\mu m}$ DG (**a**), $^{1.2\,\mu m}$ IO (**b**), $^{1.1\,\mu m}$DG (**c**), and $^{1.8\,\mu m}$ IO (**d**) electrodes

Table 5.2 Electrochromic performance at $\lambda = 430$ nm for DG and IO devices with different V_2O_5 film thicknesses

Sample	d (μm)[a]	T_c (%)	ΔT (%)	τ_c (ms)	τ_b (ms)	CCE_c (cm^2C^{-1})[b]	CCE_b (cm^2C^{-1})[b]
$^{0.73\,\mu m}$DG	0.73	16.2	52.7	71.2	53.5	−38.5	−52.5
$^{1.2\,\mu m}$IO	1.2	20.4	21.5	205.9	88.7	−28.4	−38.0
$^{1.1\,\mu m}$DG	1.1	6.1	49.8	86.7	82.5	−35.8	−33.9
$^{1.8\,\mu m}$IO	1.8	5.1	14.1	234.5	122.0	−15.6	−24.0

[a] V_2O_5 film thickness. [b] Composite coloration efficiency for a 95% switch

The ideal electrochromic material would display a large transmittance change with a small amount of charge, giving rise to a high CCE [28, 30, 33]. For the $^{1.1\,\mu m}$DG device CCE values of −33.9 and −35.8 $cm^2\,C^{-1}$ were required to achieve a 95 % switch at 430 nm, for ion insertion and extraction, respectively. This corresponds to respective CCE values of −24.0 and −15.6 $cm^2\,C^{-1}$ of the $^{1.8\,\mu m}$ IO electrode. Even higher CCE values were obtained for the thinner electrodes. These excellent CCE values are a direct consequence of the highly interconnected gyroid morphology, which combines a very high specific surface area providing good ion access with high electric conductivities.

In summary, the manufacture of a V_2O_5 in a 3D periodic highly interconnected gyroid-structure on the 10 nm length scale leads to a significant electrochromic performance enhancement. The structured devices surpass previous inorganic electrochromic materials in all relevant parameters: the switching speed, coloration contrast, and composite coloration efficiency. Compared to planar or microstructured V_2O_5, devices with the 10 nm gyroid morphology combine a high coloration contrast with a high coloration efficiency. In particular, the 85 ms switching speed lies within a factor of two of video rate and is substantially faster than the switching rates of around 1 s that are typically reported for this type of materials.

5.3.4 Electrochromic Supercapacitors Based on V_2O_5 Electrodes

Electrochemical testing was performed using supercapacitor cells shown in Fig. 5.4. The electrochemical properties of such supercapacitors with templated and nontemplated electrodes were studied by cyclic voltammetry at a sweep rate of 50 mV s^{-1} for voltage sweeps between −1 and 1 V and from −0.5 to 0.5 V. In the potential range from −0.5 to 0.5 V, both V_2O_5 materials exhibit nearly rectangular CV traces, indicating good capacitive characteristics for the devices. In the −1 to 1 V potential range, supercapacitors with lithium salt containing electrolyte (Fig. 5.14a, b) showed pseudocapacitive redox peaks. The capacitive current density of the DG V_2O_5 (Fig. 5.14a, b) is much higher compared to the significantly thicker and denser nontemplated V_2O_5 film (Fig. 5.14a).

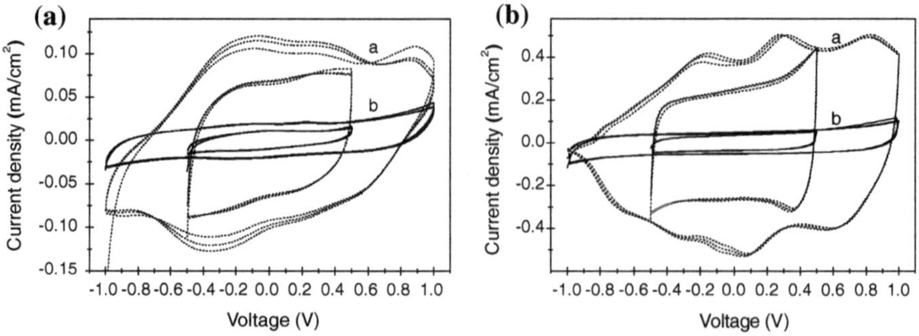

Fig. 5.14 Cyclic voltammograms of supercapacitors made with nontemplated layers (**a**) and DG-structured (**b**) electrode films, where **a** is the CV cycled with [SET3][TFSI] electrolyte containing lithium salt and **b** is the CV using Li-free RTIL as electrolyte, displaying an electric double layer response. The cycled potential ranges are -0.5 to $+0.5$ V (*inner curves*) and -1 to $+1$ V (*outer curves*) with a scan rate of 50 mV s^{-1}. Note the widely different ordinate scales

The contribution of faradaic and double-layer capacitances in the response of DG-structured V_2O_5 can be estimated by comparing the CV curves of devices that use RTIL electrolyte with and without lithium salt, shown in Fig. 5.14b. Since V_2O_5 does not react with either of the ions of pure RTIL, the lithium-free experiment tests the EDLC response. With lithium salt added, the significant increase in capacitive current and the appearance of peak pairs indicates that redox reactions are taking place. These faradaic processes are kinetically facile and thus considered pseudocapacitive, but phase transitions may occur. Although it is difficult to distinguish between redox and intercalation pseudocapacitance, the latter is likely to be present in DG bicontinuous materials.

The specific capacitance of a Li-free EDLC-based DG nanostructure is 25.8 F g^{-1}, about one order of magnitude larger compared to a nontemplated layer based device of only ~3F g^{-1}. When adding lithium salt, the specific capacitance for the DG-structured V_2O_5 pseudocapacitor raises significantly. Fig. 5.15b shows the cycle stability of the two V_2O_5-based systems containing lithium salt. The gyroid structured materials had an initial specific capacitance of 422 F g^{-1}, which equilibrated to the steady state of 155 F g^{-1} after 10 switching cycles at a high current density of >10A g^{-1}. The nontemplated layer material showed a much more gradual variation, dropping from an initial 20 F g^{-1} to 14 F g^{-1}. The charge storage ability of supercapacitors is directly linked to the geometric confinement of ions in extremely small pores and the availability of a large surface area. The much higher specific surface area of the DG-structured V_2O_5 compared to the nontemplated material therefore partly explains the substantial improvement in capacitive current.

The overall performance of our V_2O_5-based supercapacitors is summarized in the Ragone plot describing the relation between energy density and power density (Fig. 5.15a) [51]. The highest obtained energy density of nontemplated vanadia supercapacitors is 1 Wh kg^{-1} at a power density of 50 W kg^{-1}, which drops to one-half at a power density of 0.4 kW kg^{-1}. Such low values arise from the limited ion diffusion

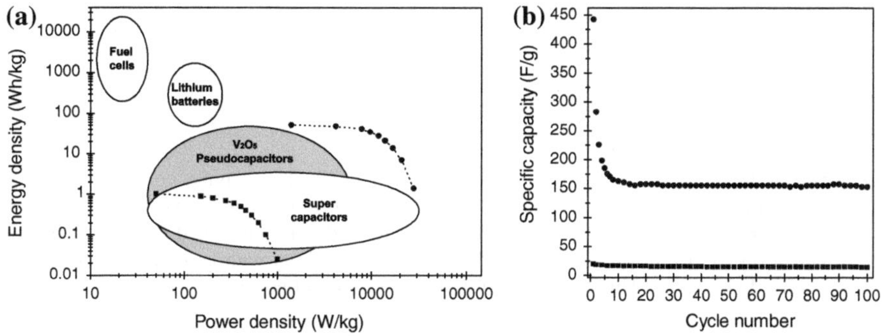

Fig. 5.15 **a** Ragone plot of electrochemical energy storage devices including the double-gyroid (•) and nontemplated layer V_2O_5 (■) supercapacitors. **b** Variation of specific capacitance with cycle number for a fixed ±4 V potential window calculated using charge-discharge curves. The double-gyroid structured pseudocapacitor gave very high initial specific capacitance of 442 F g^{-1} which stabilized at 155 F g^{-1} after 10 cycles at high current density >10 A g^{-1}. In contrast, the specific capacitance of the nontemplated V_2O_5 layer drops from an initial 20 F g^{-1} to about 14 F g^{-1}. These graphs were prepared by Dr. Di Wei

into the thick nontemplated V_2O_5 electrode, resulting also in a lack of responsiveness of the device. The maximization of the response time in nontemplated layer devices limits the thickness of most pseudocapacitive electrodes to sub-micrometer values [44, 65, 66].

Devices made from DG V_2O_5, on the other hand exhibit energy densities as high as 52 Wh kg^{-1} at power densities above 1 kW kg^{-1}. Even at power densities as high as 16.7 kW kg^{-1}, the best performing device still had an energy density of 13.9 Wh kg^{-1}, placing them among the top performing V_2O_5 devices [51]. The energy density of a supercapacitor is governed by the overall capacitance and the operating voltage. The use of organic electrolytes can increase the operating voltage compared to aqueous electrolytes [49]. RTILs are very promising solvent-free electrolytes that greatly broaden the electrochemical window during device operation, thus enabling high-voltage supercapacitors with enhanced energy densities. Combined with the introduction of interconnected porosity in DG V_2O_5, this significantly enhances the device electrochemical properties, stemming from a combination of electrolyte access, short Li^+ penetration lengths, good electronic transport, and large surface area of nanostructure enabling Li^+ insertion/extraction [44]. This leads to the observed conjunction of high specific energy and power densities of DG V_2O_5 supercapacitors, with energy densities approaching the lower end of lithium batteries.

Nontemplated V_2O_5 devices have an electrochromic switching time of several minutes and the color change is therefore not sufficiently rapid for observation when charging and discharging the device. The improvement of ion intercalation in the DG V_2O_5 supercapacitor gives rise to rapid color changes and high chromatic contrast (Figs. 5.4b and 5.16). This, the intrinsic electrochromic response, allows the visualization of the device charge state, since the degree of color saturation indicates the energy storage level of the device. In situ UV-Vis-NIR measurements of the

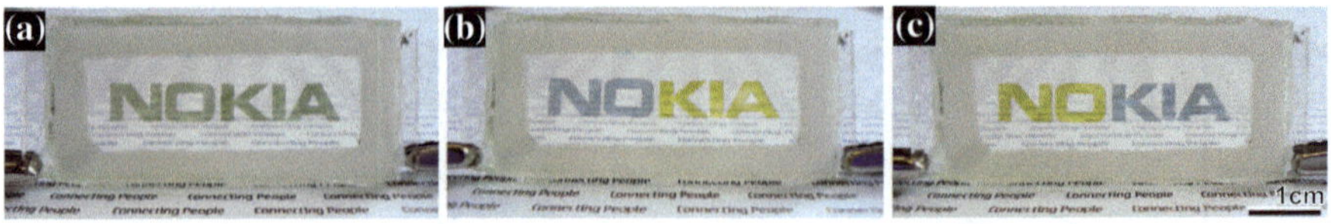

Fig. 5.16 Photographs of an electrochromic supercapacitor cell displaying color change upon charge and discharge using a potentiostat. The letters 'NO' form the double-gyroid structured top electrode and 'KIA' the bottom electrode. Both V_2O_5 electrodes have identical area. **a** Discharged *green* state at 0 V. **b**, **c** Charged state at ±3.5 V for reverse polarity with *green/gray* reduced and *yellow* oxidized electrodes. This cell is still fully functional after one year

transmissivity of DG V_2O_5 films were carried out during chronopotentiometry with a current density of $\pm 2\,\mathrm{mA\,cm^{-2}}$ and a 20 s polarity switching interval (Fig. 5.17). The anode attained the fully oxidized state, revealing a vivid yellow color, after 20 s when the potential reached 3.5 V, while the cathode exhibited the green/gray reduced state. The optical transmission spectra of the reduced and oxidized states are shown in Fig. 5.17a. The oxidized state showed a strong absorption peak around 425 nm, while the electrode was mainly transparent at wavelengths above 580 nm. The long wavelength transmittance dropped sharply during Li^+ intercalation, while rising at shorter wavelengths. The dynamic optical behavior and the potentiometric response are shown in Fig. 5.17b. The coloration change during oxidation commenced at a potential of −1 V and reached the fully oxidized state at +3.5 V, while the coloration response started at +1 V and ended at −3.5 V during reduction.

As discussed previously, Li^+ intercalation into V_2O_5 is intrinsically slow with a maximum diffusion length of approximately 5 nm for a diffusion time of 100 ms, which corresponds to the radius of the DG struts [67]. The DG-structured device therefore responds much faster than the nontemplated layer, while showing more symmetric charge-discharge curves and a much higher energy capacity, overcoming the characteristic slow ion insertion/extraction by diffusion of Li^+ through the V_2O_5 bulk. DG-structured V_2O_5 electrodes enable the reversible lithium ion

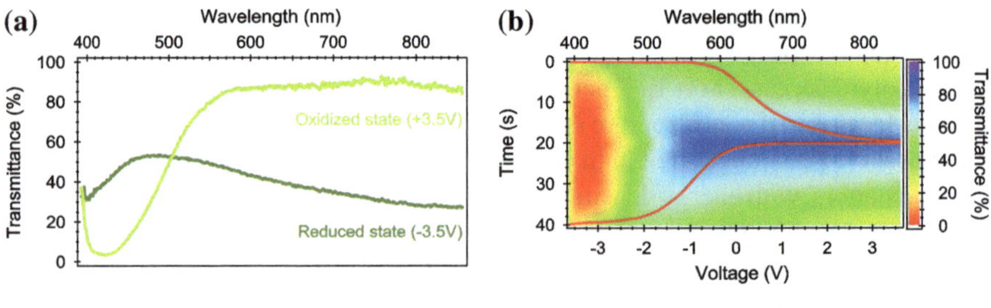

Fig. 5.17 **a** Optical transmittance spectra of the fully oxidized and reduced electrodes of the transparent supercapacitor device shown in Fig. 5.4b. **b** Transmittance variation during chronopotentiometry with alternating current polarity ($+2\,\mathrm{mA\,cm^{-2}}$ for the initial 20 s and $-2\,\mathrm{mA\,cm^{-2}}$ for the remaining time). The *red line* is the potentiometric response

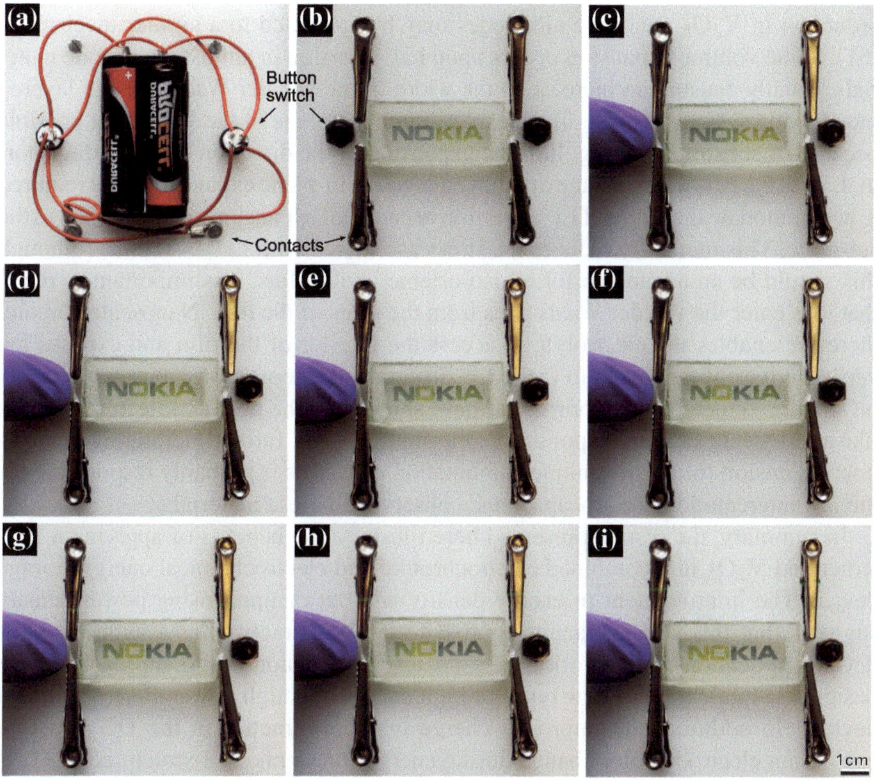

Fig. 5.18 Battery-driven stand-alone electrochromic supercapacitor device. **a** Device underside showing the wiring connecting the battery holder, the push buttons, and the clamps holding the supercapacitor cell. **b** Upside with clamps holding the supercapacitor cell from Fig. 5.16. **c–i** Pressing the *left* push button triggers a pole change of the supercapacitor cell. Likewise, pressing the *right* button reverses the process

insertion/extraction from RTIL solution in the 0–4 V potential range, corresponding to the insertion of 2 Li ions per V_2O_5 unit.

Furthermore, a battery-driven stand-alone device, see Fig. 5.18, equipped with two buttons for charging and discharging, was built. Since batteries are used rather than a potentiostat the potential and current are not dynamically adjusted, resulting in a stepwise charging/discharging on closing the circuit by pressing the push button. The color change starts at the center of the cell where the distance between the opposing V_2O_5 electrodes is minimal and gradually progresses outwards. Concerning the stability of DG-structured electrodes, this device is still fully functional after more than a year of sporadic testing.

A commonly observed problem in nanostructured supercapacitors is their initial loss of capacitance during cycling, mainly caused by compaction of the material resulting in a reduced surface area exposed to electrolyte [65]. A performance reduction in V_2O_5 nanowire electrodes may be attributed to a similar mechanism [51]. Some volume expansion occurs upon Li^+ insertion in almost all cathode materials, usually causing an increase in the width of the van der Waals gaps in layered metal oxide systems. In thin films, expansion in the plane of the film is not possible because of substrate pinning, limiting expansion in the direction of the surface normal. With the iso-oriented nature of these materials, in-plane expansion is not required to accommodate the strain of Li^+ insertion because all expansion occurs normal to the substrate. This markedly reduces the stress associated with Li^+ insertion. Although this should be an advantage for all iso-oriented bulk films, it is important to recall that ions enter the van der Waals gaps from the sides of the film. Nanoscale porosity therefore enables the electrolyte to access the interior of the film and exposes the vertical edges of the layers to facile Li^+ insertion. Moreover, previous experiments on structured metal oxide composites have shown that the pore architecture can flex like a nanoscale truss in response to a mechanical load, further facilitating this volume expansion [68]. This unique combination of factors is probably responsible for the fast intercalation pseudocapacitance observed in these materials.

In summary, the results presented here illustrate the benefits of applying a DG-structured V_2O_5 in a combined electrochromic and electrochemical energy storage device. The improvement in energy density without compromising power density suggests that this supercapacitor approach may be attractive for a wide range of device applications. This work also validates the nanocomposite approach for the design and fabrication of new types of high energy and high power electrochemical devices. In addition, the improved charge insertion kinetics in the DG structure enables an electrode color change during energy transients. DG-structured V_2O_5 is a promising candidate for high energy and high power density supercapacitors with effective color change during short-time charge and discharge cycles.

5.4 Concluding Remarks

The previously prepared voided block copolymer templates that had self-assembled into the double-gyroid morphology were successfully used for the three-dimensional nanopatterning of metal oxides, such as NiO and V_2O_5 via electrodeposition. It was demonstrated that the highly ordered structure with 11.0 nm wide struts and a high specific surface to bulk volume ratio is ideal for fast and efficient lithium ion intercalation/extraction and faradaic surface reactions, which are essential for high-performance electrochromic displays and electrochemical energy storage devices. The nanostructuring approach, as well as the fabrication of transparent supercapacitors based on an electrode material which intrinsically indicated the charge state by electrochemical color-change properties, are concepts that can be readily extended to other electrochromic intercalation compounds. However, electrodeposition of metal oxides in mesoporous templates can be sophisticated because of the fundamental deposition process involved and the poor stability of the hydrated oxides. Thus, an alternative simplified preparation route which greatly extends the catalogue of double-gyroid patternable materials is presented next. For future DG template refilling experiments yet another electrochemical method, namely the electrophoretic deposition of charged nanoparticles, might be of interest.

References

1. M.R.J. Scherer, L. Li, P.M.S. Cunha, O.A. Scherman, U. Steiner, Adv. Mater. 24(9), 1217–1221 (2012)
2. D. Wei, M.R.J. Scherer, C. Bower, P. Andrew, T. Ryhänen, U. Steiner, Nano Lett. **12**(4), 1857–1862 (2012)
3. E.J.W. Crossland, M. Kamperman, M. Nedelcu, C. Ducati, U. Wiesner, D.M. Smilgies, G.E.S. Toombes, M.A. Hillmyer, S. Ludwigs, U. Steiner, H.J. Snaith, Nano Lett. **9**(8), 2807–2812 (2009)
4. X. Zhao, B.M. Sánchez, P.J. Dobson, P.S. Grant, Nanoscale **3**(3), 839–855 (2011)
5. A.J. Bard, L.R. Faulkner, *Electrochemical Methods: Fundamentals and Applications* (Wiley, New York, 2001)
6. *On Solar Hydrogen and Nanotechnology* (Wiley, New York, 2010).
7. *Nano-Plating: Microstructure Control Theory of Plated Film and Data Base of Plated Film Microstructure* (Elsevier, London, 2004).
8. G.H.A. Therese, P.V. Kamath, Chem. Mater. **12**(5), 1195–1204 (2000)
9. T. Pauporté, A. Goux, A. Kahn-Harari, N. de Tacconi, C. Chenthamarakshan, K. Rajeshwar, D. Lincot, J. Phys, Chem. Solids **64**(9–10), 1737–1742 (2003)
10. R. Schöllhorn, Angew. Chem. Int. Ed. **19**(12), 983–1003 (1980)
11. M.S. Whittingham, Science **192**(4244), 1126–1127 (1976)
12. A.I. Khan, D. O'Hare, J. Mater. Chem. **12**(11), 3191–3198 (2002)
13. M. Vidotti, S. I. C. d. Torresi. J. Braz. Chem. Soc. **19**(7), 1248–1257 (2008)
14. K. Kang, Y.S. Meng, J. Bréger, C.P. Grey, G. Ceder, Science **311**(5763), 977–980 (2006)
15. M. Winter, J.O. Besenhard, M.E. Spahr, P. Novák, Adv. Mater. **10**(10), 725–763 (1998)
16. Y. Wang, K. Takahashi, K.H. Lee, G.Z. Cao, Adv. Funct. Mater. **16**(9), 1133–1144 (2006)
17. A. Talledo, C.G. Granqvist, J. Appl. Phys. **77**(9), 4655 (1995)
18. K. West, B. Zachau-Christiansen, T. Jacobsen, S. Skaarup, Electrochim. Acta **38**(9), 1215–1220 (1993)
19. T. Kudo, Y. Ikeda, T. Watanabe, M. Hibino, M. Miyayama, H. Abe, K. Kajita, Solid State Ionics **152–153**, 833–841 (2002)
20. A.V. Grigorieva, S.M. Badalyan, E.A. Goodilin, M.N. Rumyantseva, A.M. Gaskov, A. Birkner, Y.D. Tretyakov, Eur. J. Inorg. Chem. **2010**(33), 5247–5253 (2010)
21. Z. Lu, M. Levi, G. Salitra, Y. Gofer, E. Levi, D. Aurbach, J. Electroanal. Chem. **491**(1–2), 211–221 (2000)
22. A. Odani, V.G. Pol, S.V. Pol, M. Koltypin, A. Gedanken, D. Aurbach, Adv. Mater. **18**(11), 1431–1436 (2006)
23. D. Murphy, P. Christian, F. DiSalvo, J. Waszczak, Inorg. Chem. **18**(10), 2800–2803 (1979)
24. C.K. Chan, H. Peng, R.D. Twesten, K. Jarausch, X.F. Zhang, Y. Cui, Nano Lett. **7**(2), 490–495 (2007)
25. I. Bouessay, A. Rougier, P. Poizot, J. Moscovici, A. Michalowicz, J.-M. Tarascon, Electrochim. Acta **50**(18), 3737–3745 (2005)
26. H. Liu, G. Yan, F. Liu, Y. Zhong, B. Feng, J. Alloys Compd. **481**(1–2), 385–389 (2009)
27. S.K. Deb, Appl. Opt. 8(S1), 192–195 (1969).
28. C.G. Granqvist, *Handbook of inorganic electrochromic materials* (Elsevier, Amsterdam, The Netherlands, 1995)

29. D.R. Rosseinsky, R.J. Mortimer, Adv. Mater. **13**(11), 783–793 (2001)
30. P.M.S. Monk, P. Monk, R.J. Mortimer, D.R. Rosseinsky, *Electrochromism and Electrochromic Devices* (Cambridge University Press, Cambridge, 2007)
31. C.-G. Granqvist, Nat. Mater. **5**, 89–90 (2006)
32. S.I. Cho, W.J. Kwon, S.-J. Choi, P. Kim, S.-A. Park, J. Kim, S.J. Son, R. Xiao, S.-H. Kim, S.B. Lee, Adv. Mater. **17**(2), 171–175 (2005)
33. C.L. Gaupp, D.M. Welsh, R.D. Rauh, J.R. Reynolds, Chem. Mater. **14**(9), 3964–3970 (2002)
34. L. Li, U. Steiner, S. Mahajan, J. Mater. Chem. **20**(34), 7131 (2010)
35. C. Xiong, A.E. Aliev, B. Gnade, K.J. Balkus, ACS Nano **2**(2), 293–301 (2008)
36. K.-C. Cheng, F.-R. Chen, J.-J. Kai, Sol. Energy Mater. Sol. Cells **90**(7–8), 1156–1165 (2006)
37. K. Takahashi, Y. Wang, G. Cao, Appl. Phys. Lett. **86**(5), 053102 (2005)
38. P. Liu, S.-H. Lee, C. Tracy, J.A. Turner, J. Pitts, S.K. Deb, Solid State Ionics **165**(1–4), 223–228 (2003)
39. H. Pang, Q. Lu, Y. Zhang, Y. Li, F. Gao, Nanoscale **2**(6), 920 (2010)
40. J. Wang, L. Wei, L. Zhang, C. Jiang, E.S.-W. Kong, Y. Zhang, J. Mater. Chem. **22**(17), 8327–8335 (2012)
41. Y. Ren, W.K. Chim, S.Y. Chiam, J.Q. Huang, C. Pi, J.S. Pan, Adv. Funct. Mater. **20**(19), 3336–3342 (2010)
42. Y. Ren, S.Y. Chiam, W.K. Chim, Nanotechnology **22**(23), 235606 (2011)
43. P. Simon, Y. Gogotsi, Nat. Mater. **7**(11), 845–854 (2008)
44. T. Brezesinski, J. Wang, S.H. Tolbert, B. Dunn, Nat. Mater. **9**(2), 146–151 (2010)
45. A.S. Arico, P. Bruce, B. Scrosati, J.-M. Tarascon, W. van Schalkwijk, Nat. Mater. **4**(5), 366–377 (2005)
46. J. Wang, J. Polleux, J. Lim, B. Dunn, J. Phys. Chem. C **111**(40), 14925–14931 (2007)
47. M.J. Parent, S. Passerini, B.B. Owens, W.H. Smyrl, J. Electrochem. Soc. **146**(4), 1346–1350 (1999)
48. H. Jeon, B. Jeong, J.K. Lee, H.S. Kim, S.-H. Lee, J. Lee, Electrochem. Solid-State Lett. **13**(3), A25–A28 (2010)
49. G. Wee, H.Z. Soh, Y.L. Cheah, S.G. Mhaisalkar, M. Srinivasan, J. Mater. Chem. **20**(32), 6720–6725 (2010)
50. A. Cao, J. Hu, H. Liang, L. Wan, Angew. Chem. Int. Ed. **44**(28), 4391–4395 (2005)
51. Z. Chen, V. Augustyn, J. Wen, Y. Zhang, M. Shen, B. Dunn, Y. Lu, Adv. Mater. **23**(6), 791–795 (2011)
52. S.D. Perera, B. Patel, N. Nijem, K. Roodenko, O. Seitz, J.P. Ferraris, Y.J. Chabal, K.J. Balkus Jr, Adv. Energy Mater. **1**(5), 936–945 (2011)
53. A. Ghosh, E.J. Ra, M. Jin, H. Jeong, T.H. Kim, C. Biswas, Y.H. Lee, Adv. Funct. Mater. **21**(13), 2541–2547 (2011)
54. H. Zhang, X. Yu, P.V. Braun, Nat. Nanotechnol. **6**(5), 277–281 (2011)
55. H.Y. Lee, J. Goodenough, J. Solid State Chem. **148**(1), 81–84 (1999)
56. D. Wei, T.W. Ng, Electrochem. Commun. **11**(10), 1996–1999 (2009)
57. A.J. Meuler, C.J. Ellison, J. Qin, C.M. Evans, M.A. Hillmyer, F.S. Bates, J. Chem. Phys. **130**(23), 234903–234917 (2009)
58. E. Potiron, Electrochim. Acta **45**(1–2), 197–214 (1999)
59. W. Wang, F. Jia, Q. Huang, J. Zhang, Microelectron. Eng. **77**(3–4), 223–229 (2005)
60. Y. Wang, H. Shang, T. Chou, G. Cao, J. Phys. Chem. B **109**(22), 11361–11366 (2005)
61. X. Han, M. Tao. In Photovoltaic Specialists Conference (PVSC), 2010 35th IEEE. ISBN 0160-8371, 2010 000305-000310.
62. A.C. Sonavane, A.I. Inamdar, H.P. Deshmukh, P.S. Patil, J. Phys. D: Appl. Phys. **43**(31), 315102 (2010)
63. H. Lorenz, M. Despont, N. Fahrni, N. LaBianca, P. Renaud, P. Vettiger, J. Micromech. Microeng. **7**(3), 121–124 (1997)
64. Y.-T. Kim, S. Gopukumar, K.-B. Kim, B.-W. Cho, J. Power Sources **117**(1–2), 110–117 (2003)
65. M. Toupin, T. Brousse, D. Bélanger, Chem. Mater. **14**(9), 3946–3952 (2002)

66. S.W. Lee, N. Yabuuchi, B.M. Gallant, S. Chen, B.-S. Kim, P.T. Hammond, Y. Shao-Horn, Nat. Nanotechnol. **5**(7), 531–537 (2010)
67. S. Passerini, J. Ressler, D. Le, B. Owens, W. Smyrl, Electrochim. Acta **44**(13), 2209–2217 (1999)
68. B.L. Kirsch, X. Chen, E.K. Richman, V. Gupta, S.H. Tolbert, Adv. Funct. Mater. **15**(8), 1319–1327 (2005)

Chapter 6
Electrodeposition of Metals

After encountering a variety of difficulties during the templated electrodeposition of metal oxides and hydroxides, the aim of this part of the study was to come up with a fabrication route to DG-structured metal ceramics that avoids these shortcomings. Contradictory to intuition a two-step synthesis route involving a post-nanostructuring metal oxidation or chalcogenation can often be more versatile and economical than the direct synthesis of the desired metal oxide or chalcogenide nanostructure. Further, platinum electroplating seemed to be a suitable system for testing the theoretically predicted preferential growth directions within a gyroidal template.

Typically, electroplating metals and metal alloys is less sophisticated than the corresponding metal oxide or chalcogenide electrodeposition. Further, dense and highly stable deposits are usually obtained. The two-step synthesis approach is more versatile, in a sense that after identifying a synthesis route to a nanostructured metal, various ceramics can be produced by thermal oxidation under an appropriate atmosphere. In fact more porous and sophisticated structures, such as hollow nano-spheres, nanotubes, and nano-peapods, can be synthesized based on the nanoscale Kirkendall effect (NKE) occurring during thermal oxidation [2–5].

The NKE has been reported for numerous metals following its discovery in 2004, but most studies focused on the mentioned low-dimensional morphologies [2]. Whereas, the transformation of metals patterned in a three-dimensional, highly interconnected, periodic nanomorphology into a self-supporting nanotubular metal oxide or chalcogenide array attracted far less attention. Here, this two-step nanofabrication strategy yielding a three-dimensionally structured, nanotubular metal oxide array is tested on DG-structured Ni and Co prepared by electroplating into PS50-*b*-PLA214$|^{173\,^{\circ}\mathrm{C}}_{0.33\,\mathrm{h}}$ templates, as schematically depicted in Fig. 6.1 [6].

Parts of this chapter were previously published and are reprinted/adapted with permission from American Chemical Society, Copyright 2013 [1].

M. R. J. Scherer, *Double-Gyroid-Structured Functional Materials*, Springer Theses, DOI: 10.1007/978-3-319-00354-2_6,

Fig. 6.1 Two-step preparation of metal ceramics based on metal electroplating and subsequent thermal oxidation. Driven by the nanoscale Kirkendall effect the ceramic adopts a nanotubular DG morphology

6.1 Introduction to Metal Electroplating and Oxidation/Chalcogenation

6.1.1 Electroplating of Metals

Numerous metal and alloy coatings can be prepared by electrodeposition and are suitable for nano-plating technologies, including Ag, Au, Cd, Co, Cr, Cu, Fe, Ni, Pt, Sn, Zn, and most alloys of the listed metals [7]. Either a fairly inert counter electrode, such as a Pt mesh, a carbon rod, or a plate of the metal to be plated are used. In the latter case, the zero valent metal of the anode is oxidized and dissolved at the same rate as metal ions are reduced a the working electrode. In this manner, the cations concentration of the electrolyte bath is continuously replenished. In industrial applications usually a galvanostatic is favored over a potentiostatic deposition technique.

Unlike the metal oxide or hydroxide electrodeposition, metal plating typically only involves a simple Faradaic reaction with the direct reduction of the metal cations at the cathode to yield the desired deposit in the metallic, zero valence state,

$$M^{x+} + xe^- \rightarrow M^0. \tag{6.1}$$

This reaction mechanism which proceeds without any intermediate reactions involving other species generally favors a compact deposit and closely obeys Faraday's law, see Eq. (5.1), making a precise control over the deposition rate and thickness possible. Furthermore, the metal deposit is much more stable than the corresponding metal oxides that might be hydrated and readily redissolve. Since the metal electroplating is not considered to be a precipitation process, as in the case of various metal oxide depositions, the inclusion of impurities from the electrolyte are less likely.

Compared to the electrodeposition of the metal oxides, electroplating of the corresponding metal can have several other advantages when attempting the template-assisted nanopatterning. Metallic nickel plating, for example, is considerably faster than the deposition of nickel oxide, see Sect. 5.3.1.1. Metal plating electrolytes which are optimized to give a very homogenous growth and smooth surfaces with a bright and shiny finish are commercially available and are ideal for the replication of mesoporous templates. They come with certain additives that limit the crystal formation during electrodeposition. As such the DG-templated zinc deposition is possible, while all attempts of refilling mesoporous templates with zinc oxide failed because of the pronounced crystal formation of the oxide, see Sect. 5.3.1.1. Another advantage of metal electroplating is the ease with which coatings of well-defined alloys can be produced.

6.1.2 *Post-Nanostructuring Metal Oxidation and Chalcogenation*

Typically, thermal oxidation in microfabrication is applied to coat flat metal surfaces with a thin oxide layer, such as a silicon dioxide layer on silicon wafers. Initially, the outmost layer of metal atoms is readily converted into the corresponding oxide or chalcogenide. Any further oxidation process is then highly dependent on the interactions of metal atoms and the oxidizing agent with the formed oxidized layer. The oxidation only proceeds if the oxide is permeable either to the oxidizing agent or the metal atoms, for example, alumina is a good barrier layer that reduces the rate of any further aluminium oxidation. In the case of silicon oxidation the oxidizing agents such as water and oxygen have a higher diffusion constant than the Si atoms through the SiO_2 layer, resulting in a compact, smooth, and well-adhered oxide layer. In contrast, an inverted diffusion behavior might lead to a grainy and unstable oxide coating. However, if the metal atoms have a high mobility and the resulting oxide show a pronounced crystallization behavior, such as Cu/CuO, then the formation of a metal oxide nanowire carpet on metal surfaces is possible [8].

While the thermal oxidation of a compact metal surface is usually limited to the growth of an oxide layer with a thickness of a few of nanometers, bulk metal nanostructures can be fully converted into the corresponding oxide or chalcogenide. Again the relative diffusion rate of metal atoms and the oxidation agent in the oxide determine the oxidation kinetics and structure formation. A topographic transformation to a metal oxide nanostructure is observed when the mobility of the oxidation agent exceeds the one of the metal atoms. When this is not the case, the so-called nanoscale Kirkendall effect (NKE) responsible for the formation of sophisticated hollow nanostructures, such as nanospheres, nanotubes, and nanopeapods, proceeds [2–5].

Originally the Kirkendall effect was discovered by Smigelskas and Kirkendall in 1947 studying the interface movement resulting from non-equilibrium interdiffusion of a thermal diffusion couple in an alloying reaction [9]. It provided the first experimental proof that atomic diffusion does not occur by the direct interchange of atoms, but through vacancy exchange balancing the net directional flow of matter. In the context of metal nanostructure oxidation, the effect describes the fast outward diffusion of metal cations through the forming metal oxide/chalcogenide layer, balanced not by a considerably slower inward diffusion of chalcogen anions, but rather by an influx of vacancies [2]. The generated vacancies are forced to gradually accumulate giving rise to porosity and finally void formation. Confined by the reduced dimensions of the starting material the conversion process in most cases yields symmetric, hollow metal oxide/chalcogenide nanostructures possessing an uniform wall thickness [2–5].

Besides the fact that post-nanostructuring metal oxidation facilitates the synthesis of complex hollow morphologies it yields two further advantages. Firstly, synthesis routes such as the electrochemical replication of three-dimensional nanotemplates are often more sophisticated for metal oxides/chalcogenides than for the corresponding metals. Secondly, once a metal nanostructuring route is established the level of synthesis complexity for different oxide/chalcogenides is reduced to an optimization of the oxidation parameters, as the conversion of cobalt to CoO, CoS, $CoSe_2$, and CoTe demonstrates [2, 10]. Further, the oxidation state of the metal oxide, for example CuO and Cu_2O, can be easily controlled by adjusting the annealing temperature [11]. One requirement that has to be met by the metal and alloy nanostructure is, that the temperature necessary for the oxidation does not exceed the melting point, otherwise the nanostructure is lost. Typically, the formation of a thin oxide layer that readily forms at low temperatures increases the structural stability and makes the complete conversion to an oxide at elevated temperatures possible.

Based on the fact that Ni exhibits a much higher self-diffusion coefficient in NiO than oxygen, Nakamura et al. proposed and experimentally verified the NKE-driven void formation during thermal Ni oxidation [3, 12, 13]. However, the transformation of 50 nm Ni nanoparticles into hollow NiO nanospheres gave rise to an off-centered void position. Railsback et al. found that reducing the nano-particle diameter to 9 nm promotes the formation of nearly radially symmetric NiO shells and enables complete oxidation in the low temperature regime ($\leq$450 °C) [14]. Oxidation of Ni nanowires yielded bamboo-like nanotubes with separated voids and irregular diameters [15]. Ren et al. identified the rapid vacancy diffusion at low temperatures during the initial ramp-up stage to cause the bamboo-like structure formation during a high temperature stage (650 °C) which was necessary to completely oxidize Ni nanowires with diameters of 80 nm or larger [4, 16]. Moreover, they prepared NiO nanotubes with uniform wall thickness by replacing the high temperature oxidation step with a wet-chemical etch to remove the residual Ni metal core.

Conceptually, an ideal three-dimensional metal nanostructure for the NKE-based oxidation should be a mesoporous, fully interconnected network possessing an uniform strut diameter and sufficiently wide pore dimensions, such as the DG nanomorphology. Meeting these criteria not only prevents pore clogging during oxidative mass gain and void formation, but also gives rise to a defined formation of a self-supporting, high surface area nanotube array. Considering the studies by Railsback and Ren a Ni strut diameter of around 9 nm should facilitate a complete oxidation and a centered single void formation with a fairly uniform wall thickness in the low temperature regime [4, 14].

6.1.3 Concluding Remarks

The direct reduction of metal cations to the zero valent state at the working electrode may be advantageous for the refilling of mesoporous templates. Thus, a two-step nanofabrication strategy employing the electroplating of metals into a self-assembled mesoporous polymer template with DG morphology and the subsequent thermal oxidation might be more practical than the templated metal oxide electrodeposition. Applicable to a vast range of electroplatable metals it poses a versatile route to high surface area metal oxides/chalcogenides which is especially suitable for various thin film applications, such as electrochromic devices. Additionally, the NKE can be exploited to transform a three-dimensional, highly interconnected, periodic nanomorphology into a self-supporting nanotubular ceramic array.

6.2 Experimental Methods

6.2.1 Materials

The chemicals and materials listed in Table 6.1 were used as received.

6.2.2 Electroplating of Metals

A standard three-electrode cell equipped with an Ag/AgCl reference electrode and a platinum mesh or stainless steel counter electrode was used for the metal electroplating. Before immersing the polystyrene templates into the plating solution they were dipped into methanol to enable the infiltration of the hydrophobic mesopores by the aqueous electrolyte. Subsequently, the samples were rinsed with deionized water, dried, and electively freed from the styrenic template by dissolution in toluene for several minutes. Platinum metal was deposited potentiostatically at room

Table 6.1 Chemicals and materials used in this part of the study

Chemical	Purity [%][a]	Abbreviation
Chloroplatinic acid	99.9 (SA)	–
Cobalt sulfate	98 (AA)	–
Methanol, laboratory reagent grade	– (FS)	–
Nickel plating solution, bright finish	– (AA)	–
Oxygen	99.999 (BOC)	–
Toluene, anhydrous	99.8 (SA)	–
Water, deionized 18 MΩ	–	DI water

[a] Suppliers are given in brackets: Alfa Aesar *AA*; Fisher Scientific *FS*; Sigma Aldrich *SA*

temperature at a potential of −0.03 V. The electrolyte used was a 50 mM hexachloroplatinic acid aqueous solution. Cobalt was plated from an electrolyte containing 0.5 M cobalt sulfate as Co^{2+} source in a 0.5 M boric acid solution. Plating was performed under potentiostatic control at a constant voltage of −1.0 V at room temperature [17].

Cathodic templated and nontemplated nickel deposition was performed at a constant potential of −1.05 V at 50 °C using a commercial nickel plating solution (bright finish).

6.2.3 Post-Nanostructuring Thermal Oxidation

Nickel films were thermally oxidized in a preheated furnace at 450 °C for 10 h under an oxygen atmosphere, while cobalt nanostructures were annealed at 600 °C for 6 h [18].

6.2.4 Electrochromic Device Assembly

To investigate the electrochromic performance, transparent devices were assembled similarly to Sect. 5.2.3 by capping the prepared NiO films with a FTO counter electrode using a precut thermoplastic gasket as spacer, infiltration with 1 M $KOH_{(aq)}$ electrolyte, insertion of an Ag/AgCl wire as reference electrode, and finally sealing the device with epoxy glue.

6.3 Results and Discussion

6.3.1 Double-Gyroid-Structured Metals

6.3.1.1 Testing the Shortest Path in a Gyroidal Maze

The aim was to experimentally verify the theoretical results obtained from computer simulations on the propagation of a spherical growth front in a DG maze, which were presented in Sect. 2.4. For this purpose DG templates were prepared on untreated FTO-coated substrates that were only cleaned in an ultrasonic bath with acetone, isopropanol, and water. Omitting the surface modification step resulted in an almost complete wetting of the surface by the styrenic block. The nucleation and growth of metal was thereby limited to a few isolated pores connecting the voided DG channels with the conduction FTO layer. Templates with small and large DG domains were prepared according to the standard annealing protocol

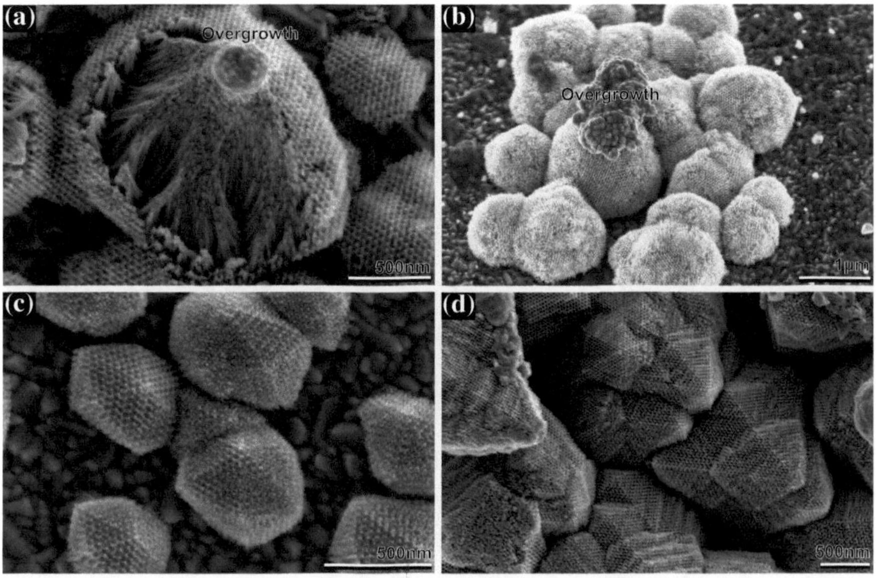

Fig. 6.2 Platinum electroplated into different DG templates. **a** Collapsed cylinders, surrounded by self-supporting DG. This nicely demonstrates why the multicontinuous DG is considered to be the superior nanostructure. Importantly, the compact, nontemplated overgrowth is spherical. **b** Replicating a P(F)S49-*b*-PLA192$|^{174\,^{\circ}\mathrm{C}}_{0.33\,\mathrm{h}}$ template with small domains results in an approximately spherical nanostructured Pt deposit with a large number of randomly orientated small-area facets. **c**, **d** Using templates of PFS45-*b*-PLA184 with fairly large single domains resulted in faceted, crystal-like deposits

using P(F)S49-*b*-PLA192$|^{174\,^{\circ}\mathrm{C}}_{0.33\,\mathrm{h}}$ and the protocol described in Fig. 4.14 from PFS45-*b*-PLA184, respectively.

Platinum was chosen as the metal to be electroplated, since a nontemplated deposit shows a compact and spherical growth, an assumption required by the theoretical computer model, see Fig. 6.2a. From this it was inferred, that electroplated platinum is highly polycrystalline, but this still has to be confirmed by electron diffraction. In contrast, metals such as tin and copper readily crystallize during electrodeposition and form faceted crystals. As predicted by the computer model, the spherical growth front of highly polycrystalline platinum in the voided, ‘single-crystalline’ gyroidal maze results in faceted, crystal-like deposits, as presented in Fig. 6.2c, d. This delivers an experimental proof that the template’s symmetry can be passed onto an amorphous or highly polycrystalline replica by adopting a crystal-like shape that reflects the template’s symmetry. An additional control experiment was conducted using a highly polycrystalline template with small domains. Due to the lack of long-range order the replica are not expected to show a pronounced crystal-like appearance. Indeed, more spherical platinum deposits with a high number of small facets were observed when refilling these templates, see Fig. 6.2b.

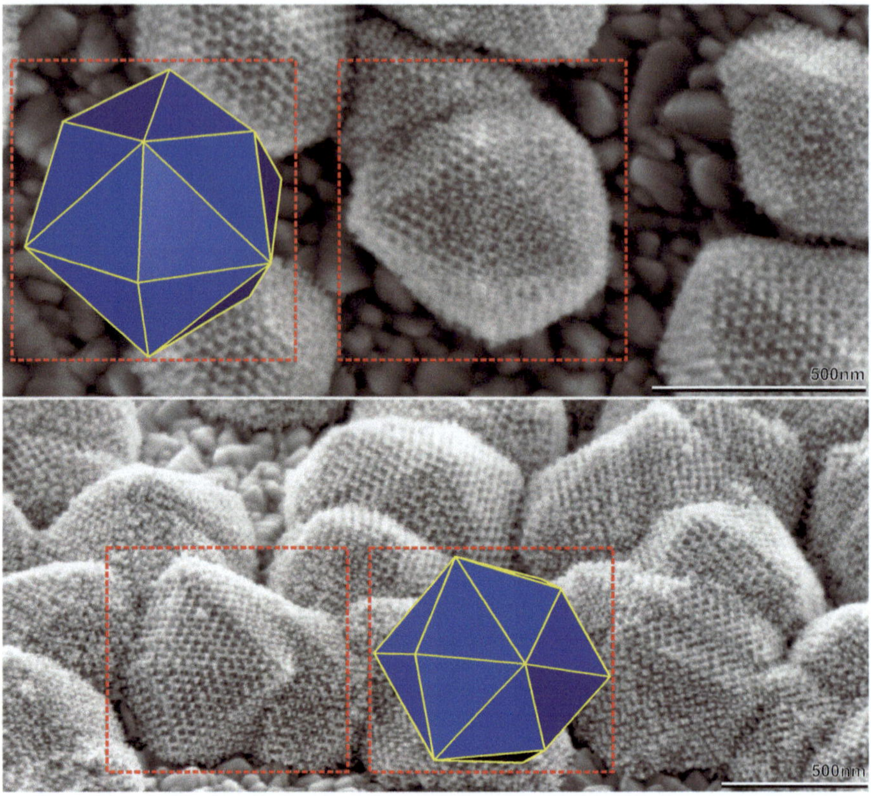

Fig. 6.3 SEM images of crystal-shaped platinum deposits juxtaposed to computer simulations. The *dashed red frame* marks the area of interest

Figure 6.3 shows crystal-shaped platinum deposits juxtaposed to the computer simulations. Note that the computer simulation were calculated for a minority phase volume fraction of 12 %, while depending on the assumed polymer densities, the copolymer PLA volume fraction lies between 36 and 39 %. As discussed previously, for these high volume fractions the computer model predicts that the <110> directions form vertices and thus, increase the number of faces surrounding the <100> vertices by a factor of two to a total of 8 facets.

Two plausible reasons might explain why the simulations obtained for smaller PLA volume fractions provide a good fit to the experimental results. Firstly, as argued in Sect. 2.4, due to the shallowness of the angles between the four faces adjoining at a <110> vertex they may appear as one larger face instead of four individual faces when imaged with an electron microscope. Secondly, the double-gyroid platinum replica appears to be more sparse than expected for a network volume fraction of 36–39 %. Comparing the porosity of simulated double-gyroid networks for different volume fractions to the SEM images suggests a volume fraction well below 30 %. This discrepancy will be studied in more detail.

Nevertheless, aligning the SEM electron beam with the direction of a vertex surrounded by six faces reveals that as predicted this vertex coincides with the <111> direction. Similarly, vertices surrounded by four faces are consistent with the <100> direction.

6.3.1.2 Electroplating of Continuous Metallic Films

Whereas templates with a low porosity at the film/substrate interface were intentionally used above to limit the number of nucleation sites, here fully porous templates were used which facilitated a homogenous and dense nucleation across the working electrode. This was achieved by modifying the FTO surface chemistry by piranha etching and the deposition of an imperfect OTS monolayer, which tailored a neutral surface for both copolymer blocks. Due to a high nucleation density the crystal formation based on the preferential growth directions in the gyroidal template was suppressed, but since PS50-*b*-PLA214$|_{0.33\,\mathrm{h}}^{173\,^{\circ}\mathrm{C}}$ scaffolds consist of many domains, some facets were formed at the free-surface. Overall, the deposited metal film thickness was very homogeneous across the deposition area and the free-surface was fairly smooth, see Fig. 6.4. This was true for all test metals. Importantly, the silane which should act as an insulating barrier layer, does not prevent the deposition, most probably due to defects and pinholes in the imperfect monolayer. In fact the silane seems to promote the adhesion of the metallic deposit to the substrate.

For the study on the oxidation of double-gyroid metal films and the subsequent use of NiO films in electrochromic devices, templated and nontemplated metallic films of different thickness were prepared. An uniform Ni deposition thickness, which is important when using the oxidized electrodes in electrochromic cells was achieved by limiting the plating area to 1 mm^2 using a SU-8 mask [19].

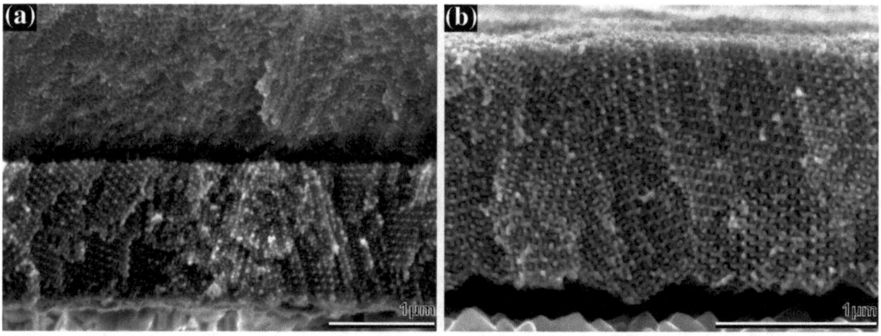

Fig. 6.4 Cross-sectional SEM images of templated nickel films. **a** Partially refilled template. **b** Free-standing nickel array imaged after template removal revealing a strut diameter of 11.0 ± 0.3 nm

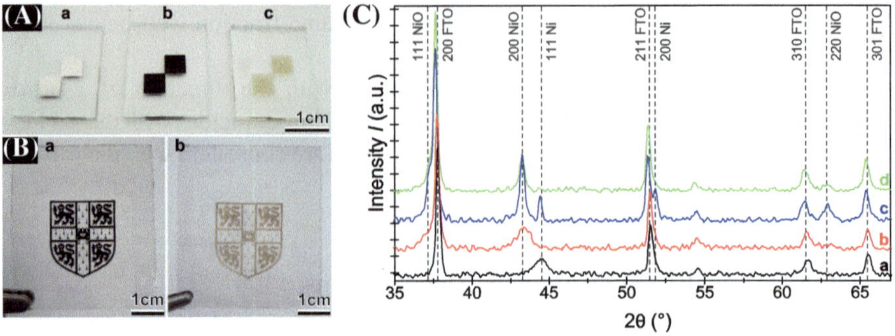

Fig. 6.5 **a** Photograph of an as-deposited nontemplated Ni film showing a bright mirror-like finish (*a*), while the as-deposited DG Ni film is pitch black (*b*) that turns semi-transparent when oxidized (*c*). **b** As-deposited (*a*) and thermally oxidized (*b*) templated nickel pattern. **c** XRD pattern (Cu K_{α} radiation) for an as-deposited DG (*a*), a thermally annealed 520 nm thick DG (*b*), and two thermally annealed nontemplated films with a thickness of 210 and 70 nm (*c* and *d*) on FTO-coated glass. The observed indexed peaks are matched with the corresponding tetragonal SnO_2 (PDF number: 046-1088), cubic Ni (PDF number: 004-0850), and cubic NiO (PDF number: 047-1049). While the thinner nontemplated nickel deposit was fully oxidized, the nickel peaks are still present for the thicker film

The nickel deposition rate was found to be $0.288\,\mathrm{mg\,C^{-1}}$, which corresponds to 32.4 and 85.4 nm per $\mathrm{mC\,mm^{-2}}$ for a nontemplated and templated film, respectively. This value is close to the theoretical upper limit set by Faraday's law and translates into a good current efficiency. Micron-thick nickel films were plated within minutes while the deposition of comparable nickel oxide coatings took 100 min, see Sect. 5.3.1.1. In contrast to the bright, mirror-like finish of nontemplated Ni or Co deposits, gyroid-structured metallic films are highly light absorbing resulting in a pitch black appearance, see Fig. 6.5a. This vividly demonstrates the profound impact structural manipulation on the nanometer scale has on material properties. Magneto-optical Kerr effect measurements on double-gyroid-structured Ni and Co films, which were performed by members of the Thin Film Magnetism Group at the Cavendish Laboratory, showed an unique behavior, that is not observed for nontemplated films. However, since the author was not involved in the measurements no results are presented here.

6.3.2 Thermal Oxidation of Double-Gyroid-Structured Metals

Ni films were thermally oxidized in a preheated furnace at 450 °C for 10 h under an oxygen atmosphere resulting in a brownish, semi-transparent appearance (Fig. 6.5a, b). X-ray diffraction (XRD) revealed that this treatment was sufficient to thoroughly oxidize a 70 nm thick nontemplated film as well as templated Ni of arbitrary thickness, while a 210 nm thick nontemplated deposit was only partially converted to

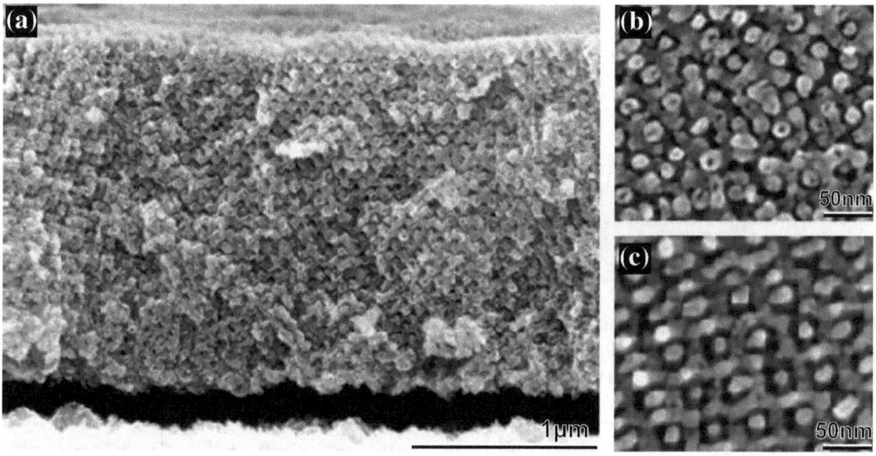

Fig. 6.6 SEM images of thermally oxidized NiO with a hollow DG morphology based on the Kirkendall mechanism. **a** Cross-section of a NiO nanotube array obtained by thermal annealing subsequent to template dissolution. **b** Free-surface of the film shown in (**a**) with pronounced nanotube formation. **c** Free-surface of a templated nickel film which was oxidized with the styrenic template present. The template confines the oxidizing Ni struts and thereby, the nanotube void formation

NiO (Fig. 6.5c). In the latter sample, both nickel reflection peaks were still present. Comparing the 200 peak width of the templated and nontemplated samples reveals that the DG NiO is more polycrystalline. In case of the nanopatterned nickel with a strut diameter of only 11.0 ± 0.3 nm, a significantly shorter annealing duration would probably have been sufficient to achieve complete oxidation.

SEM images of a nanotubular nickel oxide DG obtained after thermal annealing based on the Kirkendall effect is shown in Fig. 6.6. The extent of void formation driven by the Kirkendall effect can be controlled by the elective dissolution of the template before thermal annealing. Thermal annealing performed on template-freed samples led to an unhindered expansion of the Ni struts which is based on the NKE with pronounced nanotube formation (Fig. 6.6b). In contrast, the expansion of the oxidizing Ni struts as well as the extent of nanotube void formation is constrained (Fig. 6.6c). The following discussion focuses on nanotubular NiO with pronounced void formation. The nanotubular array can be considered as a core-shell double-gyroid which was discussed in Sect. 2.6.

Transmission electron micrographs (TEMs) confirmed the porosity of the NiO nanotubes and further revealed their substructure to be nanocrystalline with crystallite sizes ranging from 5 to 10 nm, see Fig. 6.7. The oxidation of nanopatterned cobalt films worked equally well [18]. The gyroid structure was nicely preserved during oxidation. The XRD data present in Fig. 6.8 suggests a complete conversion to Co_3O_4. Nanostructured cobalt oxide is a battery electrode material, but due to time constrains this was not tested in this study.

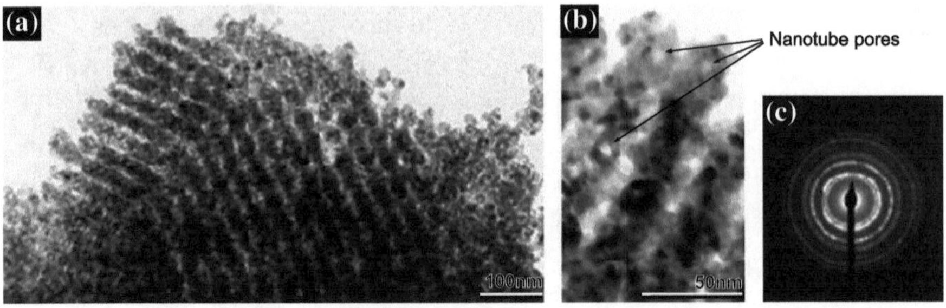

Fig. 6.7 **a** TEM revealing the nanotubular character and polycrystallinity of the double-gyroid-structured NiO. **b** Magnified view of (**a**) resolving nanocrystals with dimensions of 5–10 nm. **c** Corresponding selected area electron diffraction pattern of polycrystalline NiO.

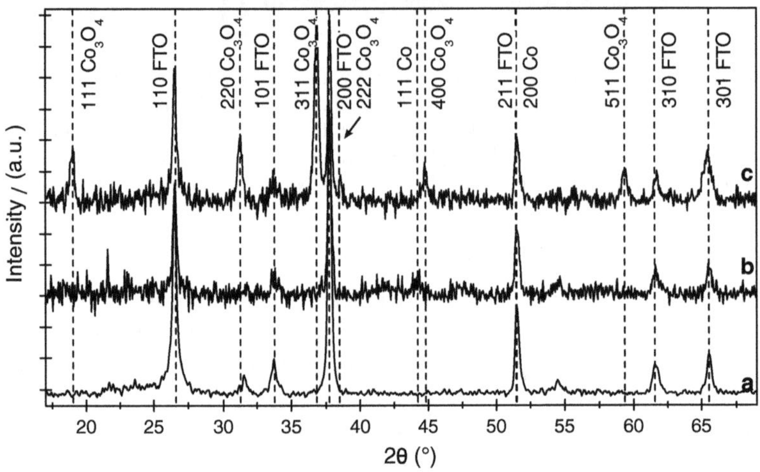

Fig. 6.8 XRD patterns of a bare FTO substrate (*a*), a DG-structured metallic (*b*) and thermally oxidized (*c*) cobalt film annealed at 600 °C for 6 h. Cu K_{α} radiation was used and the patterns were matched with the corresponding tetragonal SnO_2 (PDF number: 046-1088), cubic Co (PDF number: 015-0806), and cubic Co_3O_4 (PDF number: 042-1467)

6.3.3 Electrochromic Displays Based on NiO Electrodes

As naturally abundant and low-cost semiconductor, NiO is widely used in electrochromic windows [20], batteries [21], supercapacitors [22], and sensors [23]. While all these applications benefit from an interconnected, three-dimensional NiO nanostructure that combines a high specific surface area with a good electric conductivity, the performance enhancement becomes vividly evident as an increased coloration contrast and improved switching behavior when applied in electrochromic devices. NiO nanomaterials recently employed in electrochromic studies include nanocomposites [24], inverse opals [25], macroporous [26] and mesoporous films [27–29].

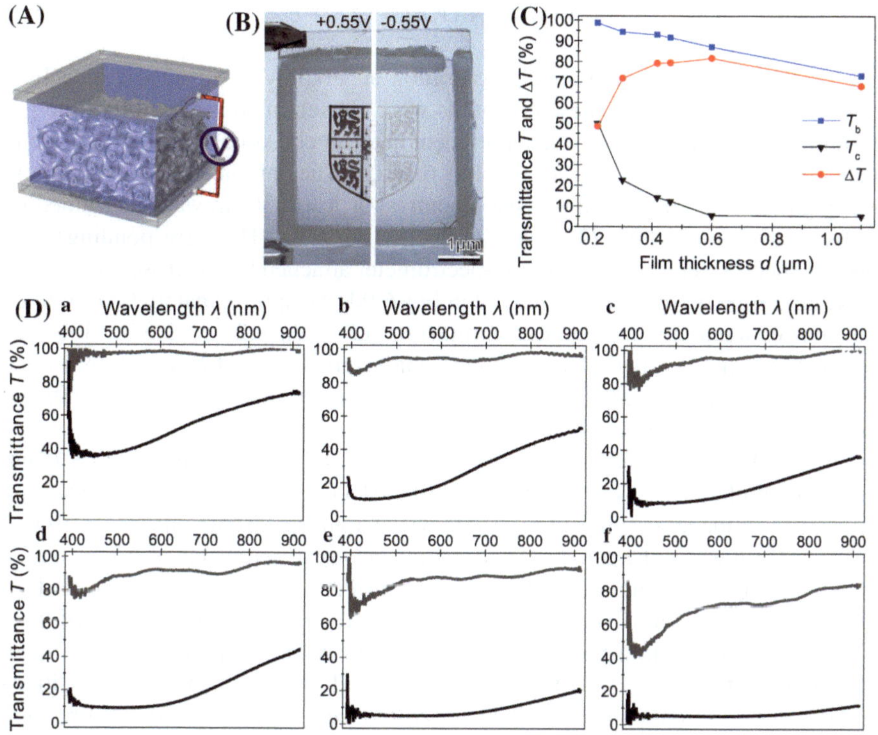

Fig. 6.9 **A** Schematic illustrating the employed device design based on a DG NiO nanotube array. **B** Transparent electrochromic device consisting of a FTO/NiO/$KOH_{(aq)}$/FTO layer sequence and an Ag/AgCl wire as reference electrode. The design pattern in the chromogenic DG film is the crest of the University of Cambridge, showing the colored oxidized (NiOOH) and the bleached reduced ($Ni(OH)_2$) state. **C** Dependence of the optical transmission at $\lambda = 630$ nm on the film thickness of templated NiO in the colored (T_c at +0.55 V) and bleached (T_b at −0.55 V) state tested with 1 M $KOH_{(aq)}$. A ≈ 600 nm thick film showed a maximum coloration contrast of $\Delta T = T_b - T_c = 81.5\,\%$. **D** Corresponding transmission spectra for templated NiO electrodes with the following film thicknesses: **a** 215 nm, **b** 300 nm, **c** 416 nm, **d** 460 nm, **e** 600 nm, and **f** 1,100 nm.

To investigate the effect of the nanostructure on the electrochromic performance of NiO, transparent devices were assembled from nontemplated and DG-structured films with a FTO counter electrode and a 1 M $KOH_{(aq)}$ electrolyte, see Fig. 6.9a. The active electrode material used was limited in area to 0.95 mm^2. During the nickel electroplating process limiting the deposition area improve the control and quality of the deposit.

In alkaline solution the reversible optical absorption change caused by an electron-transfer reaction accompanied by a compensating transfer of mobile H^+ cations between the hydrated NiO and electrolyte can be summarized by the following reaction scheme [30, 31]

$$Ni(OH)_2 \rightleftarrows NiOOH + H^+ + e^-. \tag{6.2}$$

The NiO samples exhibited anodic coloration at +0.55 V and cathodic bleaching at −0.55 V, see Fig. 6.9b. These potentials were chosen in consideration of the gas evolution onset at higher potential, as evident in the cyclic voltammograms as strong increase in the current, as displayed in Fig. 6.10d, e. The optimal DG NiO film thickness regarding the optical transmission contrast at $\lambda = 630$ nm was determined by analyzing six samples after 750 switching cycles (Fig. 6.9c). The corresponding transmission spectra were recorded with a spectrometer attached to a microscope with the integration time set to 25 ms, displayed in Fig. 6.9d. An approximately 600 nm thick film showed a maximum coloration contrast of $\Delta T = T_b - T_c = 81.5\,\%$. However, for the following detailed electrochromic characterization a thinner, 460 nm thick DG electrode was chosen, since it showed a similar coloration contrast combined with a superior coloration efficiency.

Figure 6.10a shows the spectral response of the DG in comparison with a 70 nm thick nontemplated film. The color contrast of the DG benefits from a pronounced and uniform variation across the entire visible wavelength spectrum with $\Delta T = 79.6\,\%$ at $\lambda = 630$ nm where the nontemplated sample only reaches a value of $\Delta T = 6.3\,\%$. At $\lambda = 580$ nm, a maximum coloration contrast of $\Delta T = 82.2\,\%$ was measured for the DG device, in comparison to $\Delta T = 7.7\,\%$ for a 70 nm thick nontemplated film.

A second important aspect of electrochromism is the temporal response under alternating potentials (±0.55 V). The DG showed sharp and distinct transitions between the colored/oxidized and bleached/reduced state across the entire visible spectrum (Fig. 6.10b). This time-resolved switching behavior was analyzed in more detail at $\lambda = 630$ nm (Fig. 6.10c). The DG device showed short characteristic response times of 53 ms for the bleaching step and 63 ms for the reverse process, determined by fitting exponential functions to the switching curves. These short response times which are close to video rate (24 frames per second) are enabled by the short ion diffusion distance through the only ≈ 5 nm thick NiO nanotube wall.

The composite coloration efficiency $CCE = \Delta OD/q$ is defined by the charge per area q that is required to excite a gradient in optical density $\Delta OD = \ln(T/T_0)$, where T_0 is the initial transmittance (Fig. 6.10f). The ideal electrochromic material should maximize its CCE, corresponding to a large transmittance change induced by a small amount of charge [32–34]. For the DG NiO morphology an average coloration efficiency at $\lambda = 630$ nm of −47 and $-40\,\mathrm{cm^2\,C^{-1}}$ were achieved for the 10th and 750th switching cycle, respectively (Fig. 6.10f). These values are consistent with those reported for NiO prepared by electrodeposition, chemical bath and chemical vapor deposition [25, 35, 36]. These excellent *CCE* values in combination with a high $\Delta OD = 2.1$ are a direct consequence of the highly interconnected gyroid morphology, which combines a very high specific surface area providing good ion access with a high electric conductivity.

Cycle stability and stability of the structural integrity of the nanostructured material are further important requirements of electrochromic materials. The CV curves presented in Fig. 6.10d, e indicate that the nanotubular NiO structures exhibits good long-term cycle stability, giving rise to a pair of well pronounced redox peaks. Fur-

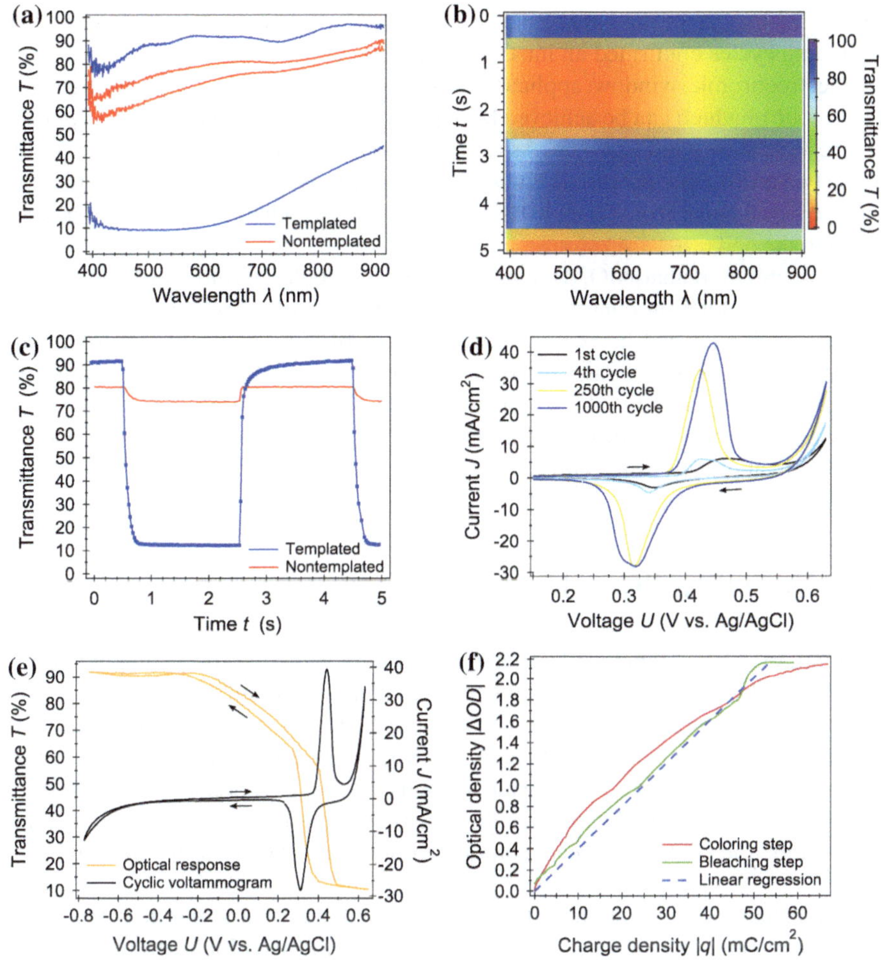

Fig. 6.10 Electrochromic performance of a 460 nm thick DG-nanopatterned NiO film in 1 M $KOH_{(aq)}$ electrolyte after 750 full switching cycles. **a** Optical transmittance spectra of the colored (+0.55 V) and bleached (−0.55 V) state. **b** Transmittance variation upon alternating 2 s potential steps (±0.55 V) recorded with a 25 ms integration time visualizing a consistently fast switching response over the entire visible spectrum. **c** Corresponding switching behavior at $\lambda = 630$ nm. The DG device showed a maximum coloration contrast of $\Delta T = 79.6\,\%$ and fitting the data with $T(t) = T_0 \pm \Delta T \cdot e^{-t/\tau}$ gave characteristic switching times τ of 63 and 53 ms for the coloring and bleaching process, respectively. **d** CVs recorded at a scan rate of 50 mV s^{-1}. After an initial activation period the two redox peaks become more pronounced with increasing cycle number. The current increase at high potentials is caused by the onset of gas evolution. **e** Optical transmittance response at $\lambda = 630$ nm as function of the applied cell potential and corresponding CV recorded for the 750th switching cycle at a 50 mV s^{-1} scan rate. **f** Optical density $|\Delta OD|$ at $\lambda = 630$ nm versus the charge density $|q|$ calculated from the data presented in **E**. The linear regression suggests an average coloration efficiency of $-40\,cm^2\,C^{-1}$

ther, SEM images of the NiO electrode after electrochromic testing including 1000 switching cycles confirmed an intact nanotubular nanostructure.

Electrochromic window applications may require a more adjustable transmittance control which can be achieved by applying a cell potential between −0.55 and +0.55 V, as shown in Fig. 6.10e. As expected a steep change in transmittance occurs at the two redox peak potentials giving rise to hysteresis. However, the transmittance change in the range from 60 to 90 % is more gradual with hardly any current passing through the cell.

Compared to planar NiO, devices with the nanotubular gyroid morphology combine a high coloration contrast with a high coloration efficiency. In particular, the 63 ms switching speed lies within a factor of 1.5 of video rate and is substantially faster than the switching rates of around 1 s that are typically reported for this type of materials. Importantly, the template forming copolymer used in this part of the study consists only of cheap, and environmentally sustainable monomers, making large scale applications feasible.

6.4 Concluding Remarks

In summary, gyroid templated metal electroplating was used for testing the preferential growth directions in gyroidal templates with long-range order and to prepare samples for the thermal oxidation of three-dimensionally nanopatterned metals. In particular, it was shown that the nanoscale Kirkendall effect can be utilized to prepare sophisticated hollow metal oxide nanotube arrays. NiO prepared in this fashion was found to significantly enhance the electrochromic performance in all relevant parameters: the switching speed, coloration contrast, and composite coloration efficiency.

The enhanced material properties of the nanotubular double-gyroid morphology can be extended to other electroplatable transition-metals which exhibit the Kirkendall mechanism and is therefore promising for batteries, supercapacitors, and sensors. So far, only oxygen was tested as oxidizing agent. In order to prepare metal sulfides by thermal oxidation, hydrogen sulfide could be used as a sulfur source [37]. For solar cell applications, stibnite or $CuInS_2$ with double-gyroid morphology might be highly interesting [38–41]. The electrodeposition of antimony, the thermal oxidation to Sb_2O_3, as well as the sulfidation of the oxide to Sb_2S_3 have been reported [42–44]. Further, electroplating and oxidation of nanopatterned tin, copper, or indium alloys could find widespread applications as transparent a conducting electrode material [11, 45–49].

References

1. M.R.J. Scherer, U. Steiner, Nano Lett. **6**, 2325–2331 (2013)
2. Y. Yin, R.M. Rioux, C.K. Erdonmez, S. Hughes, G.A. Somorjai, A.P. Alivisatos, Science **304**(5671), 711–714 (2004)
3. R. Nakamura, J.-G. Lee, H. Mori, H. Nakajima, Philos. Mag. **88**(2), 257–264 (2008)
4. Y. Ren, W.K. Chim, S.Y. Chiam, J.Q. Huang, C. Pi, J.S. Pan, Adv. Funct. Mater. **20**(19), 3336–3342 (2010)
5. Y. Yang, L. Liu, F. Güder, A. Berger, R. Scholz, O. Albrecht, M. Zacharias, Angew. Chem. Int. Ed. **50**(46), 10855–10858 (2011)
6. H.-Y. Hsueh, Y.-C. Huang, R.-M. Ho, C.-H. Lai, T. Makida, H. Hasegawa, Adv. Mater. **23**(27), 3041–3046 (2011)
7. T. Watanabe, *Nano-Plating: Microstructure Control Theory of Plated Film and Data Base of Plated Film Microstructure* (Elsevier, London, 2004)
8. J. Liang, N. Kishi, T. Soga, T. Jimbo, J. Nanomater. **2011**, 1–8 (2011)
9. A.D. Smigelskas, E.O. Kirkendall, Trans. AIME. **171**, 130–142 (1947)
10. J. Gao, B. Zhang, X. Zhang, B. Xu, Angew. Chem. **118**(8), 1242–1245 (2006)
11. J. Park, K. Natesan, Oxid. Met. **39**(5), 411–435 (1993)
12. A. Atkinson, R.I. Taylor, Philos. Mag. A **39**(5), 581–595 (1979)
13. M. O'Keeffe, W.J. Moore, J. Phys. Chem. **65**(8), 1438–1439 (1961)
14. J.G. Railsback, A.C. Johnston-Peck, J. Wang, J.B. Tracy, ACS Nano **4**(4), 1913–1920 (2010)
15. R. Nakamura, G. Matsubayashi, H. Tsuchiya, S. Fujimoto, H. Nakajima, Acta Mater. **57**(17), 5046–5052 (2009)
16. Y. Ren, S.Y. Chiam, W.K. Chim, Nanotechnology **22**(23), 235606 (2011)
17. A. Llavona, C. Díaz-Guerra, M. Sánchez, L. Perez, Mater. Chem. Phys. **124**(2–3), 1177–1181 (2010)
18. T. Li, S. Yang, L. Huang, B. Gu, Y. Du, Synthesis of antimony oxide nanoparticles by thermal oxidation **15**(11), 1479–1482 (2004)
19. H. Lorenz, M. Despont, N. Fahrni, N. LaBianca, P. Renaud, P. Vettiger, J. Micromech. Microeng. **7**(3), 121–124 (1997)
20. G.A. Niklasson, C.G. Granqvist, J. Mater. Chem. **17**(2), 127 (2007)
21. A. Shukla, S. Venugopalan, B. Hariprakash, J. Power Sources **100**(1–2), 125–148 (2001)
22. K.-W. Nam, K.-B. Kim, J. Electrochem. Soc. **149**(3), A346 (2002)
23. J.A. Dirksen, K. Duval, T.A. Ring, Sens Actuators. B **80**(2), 106–115 (2001)
24. S.-H. Lin, F.-R. Chen, J.-J. Kai, Appl. Surf. Sci. **254**(11), 3357–3363 (2008)
25. Y. Yuan, X. Xia, J. Wu, Y. Chen, J. Yang, S. Guo, Electrochim Acta **56**(3), 1208–1212 (2011)
26. X. Xia, J. Tu, J. Zhang, X. Wang, W. Zhang, H. Huang, Sol. Energy Mater. Sol. Cells **92**(6), 628–633 (2008)
27. A.C. Sonavane, A.I. Inamdar, H.P. Deshmukh, P.S. Patil, J. Phys. D: Appl. Phys. **43**(31), 315102 (2010)
28. D. Dalavi, M. Suryavanshi, D. Patil, S. Mali, A. Moholkar, S. Kalagi, S. Vanalkar, S. Kang, J. Kim, P. Patil, Appl. Surf. Sci. **257**(7), 2647–2656 (2011)
29. D.S. Dalavi, M.J. Suryavanshi, S.S. Mali, D.S. Patil, P.S. Patil, J. Solid State Electrochem. **16**(1), 253–263 (2011)
30. I. Bouessay, A. Rougier, P. Poizot, J. Moscovici, A. Michalowicz, J.-M. Tarascon, Electrochim. Acta **50**(18), 3737–3745 (2005)
31. H. Liu, G. Yan, F. Liu, Y. Zhong, B. Feng, J. Alloys Compd. **481**(1–2), 385–389 (2009)
32. P.M.S. Monk, P. Monk, R.J. Mortimer, D.R. Rosseinsky, *Electrochromism and Electrochromic Devices* (Cambridge University Press, Cambridge, 2007)
33. C.G. Granqvist, *Handbook of Inorganic Electrochromic Materials* (Elsevier, Amsterdam, The Netherlands, 1995)
34. C.L. Gaupp, D.M. Welsh, R.D. Rauh, J.R. Reynolds, Chem. Mater. **14**(9), 3964–3970 (2002)
35. X. Xia, J. Tu, J. Zhang, X. Huang, X. Wang, W. Zhang, H. Huang, Electrochem. Commun. **10**(11), 1815–1818 (2008)

36. T. Maruyama, S. Arai, Sol. Energy Mater. Sol. Cells **30**(3), 257–262 (1993)
37. W. Qian, A. Ishihara, Y. Aoyama, T. Kabe, Appl. Catal., A **196**(1), 103–110 (2000)
38. M. Nanu, L. Reijnen, B. Meester, J. Schoonman, A. Goossens, Chem. Vap. Deposition **10**(1), 45–49 (2004)
39. S.-J. Moon, Y. Itzhaik, J.-H. Yum, S.M. Zakeeruddin, G. Hodes, M. GrStzel, J. Phys. Chem. Lett. **1**(10), 1524–1527 (2010)
40. I. Mora-Sero, J. Bisquert, J. Phys. Chem. Lett. **1**(20), 3046–3052 (2010)
41. T. Dittrich, A. Belaidi, A. Ennaoui, Sol. Energy Mater. Sol. Cells **95**(6), 1527–1536 (2011)
42. Y. Sadana, J. Singh, R. Kumar, Surf. Technol. **24**(4), 319–353 (1985)
43. C.H. Xu, S.Q. Shi, Q. Tang, Nanotech **2**, 54–57 (2005)
44. B. Juárez, M. Ibisate, J. Palacios, C. López, Adv. Mater. **15**(4), 319–323 (2003)
45. A. Kolmakov, Y. Zhang, G. Cheng, M. Moskovits, Adv. Mater. **15**(12), 997–1000 (2003)
46. A. Kolmakov, Y. Zhang, M. Moskovits, Nano Lett. **3**(8), 1125–1129 (2003)
47. B. Luo, D. Yang, M. Liang, L. Zhi, Nanoscale **2**(9), 1661–1664 (2010)
48. X. Li, E. Chin, H. Sun, P. Kurup, Z. Gu, Sens. Actuators, B **148**(2), 404–412 (2010)
49. A. He, Q. Liu, D.G. Ivey, J. Mater. Sci. Mater. Electron. **19**(6), 553–562 (2007)

Chapter 7
Electropolymerization of Conjugated Polymers

This chapter investigates the nanostructuration of intrinsically conducting polymers via templated electropolymerization using polymeric scaffolds with a double-gyroid morphology, see Fig. 7.1. Conjugated polymers combine some of the functional properties of inorganic semiconductors and/or metals with the favorable material properties of insolating polymers, such as flexibility and processability. The nanopatterned conjugated polymers demonstrated in this chapter include poly(pyrrole), poly(3,4-ethylenedioxythiophene), poly(thiophene), and poly(3-methylthiophene) denoted as PPy, PEDOT, PT, and P3MT, respectively. The performances of these polymers in a variety of applications are expected to be greatly enhanced by the well-ordered and highly porous nanostructure which exhibits a very large surface area [1–6].

So far, only the DG-structured poly(pyrrole) films were tested in functional devices. Applied in chemical gas sensors, these films showed a sensitive response to the presence of solvent vapors, but their performance was not significantly improved compared to their nontemplated counterparts. This topic will therefore only be briefly discussed here and the reader is referred to the excellent thesis of **Alexandre Nicolas**, who worked on this topic as part of his MPhil project under my supervision [7]. Although the other templated films have only been partly characterized, regarding their chemical, electronic, and optical properties in the doped and undoped state, we believe that they bear the potential to gain importance as active electrodes material in batteries, supercapacitors, and photovoltaic devices [8–11]. The latter devices may be prepared by infiltrating dedoped, mesoporous PT or P3MT films with [6,6]-phenyl-C_{61}-butyric acid methyl ester (PCBM) [12–15]. First encouraging results were very recently obtained by **Harry Beeson** and **Micheal Price**, who characterized these double-gyroid-structured thiophenic polymer films regarding their application in bulk heterojunction solar cells under my supervision.

The experimental implementation proved to be challenging, not least because of the fact, that both the template and the replicated material are organic and therefore subject to dissolution and swelling on exposure to organic solvents. Several different orthogonal solvents had to be identified for the consecutive fabrication steps, including suitable solvents for the deposition bath, the template dissolution,

M. R. J. Scherer, *Double-Gyroid-Structured Functional Materials*,
Springer Theses, DOI: 10.1007/978-3-319-00354-2_7,

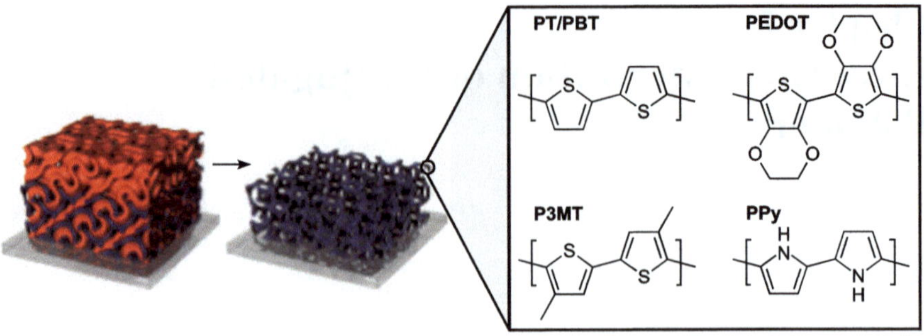

Fig. 7.1 Schematic illustrating the template-assisted nanopatterning of conjugated polymers via electrodeposition

the dedoping of the mesoporous deposit, and where applicable the refilling of the latter with other functional materials. Although a number of solvents are considered 'non-solvents' for (substituted) poly(styrene) and for conjugated polymers, which are typically known to be insoluble in most common solvents, nanostructures of all these polymers were found to be highly susceptible to swelling by some of these 'non-solvents'. Typically, the polymeric double-gyroid films showed fair structural stability when exposed to mechanical stresses, but the swelling by organic liquids (including 'non-solvents') resulted in pore clogging, collapse of the nanostructure, and/or cracking. However, with the help of solubility parameters, suitable solvents were identified and successfully applied during the multiple fabrication steps.

7.1 Introduction to Conjugated Polymers, Electropolymerization, and Solubility Parameters

7.1.1 Conjugated Polymers

As high-efficiency organic conductors, conjugated polymers have attracted much attention since the discovery of conducting poly(acetylene) by Shirakawa et al. in 1977, but reports concerning the electrical properties of polymers date back at least to the 1960s [16, 17]. The difference in electric properties between insulating and intrinsically conducting polymers arises from the different hybridization of the carbon atoms forming the polymer backbone.

The carbon atoms of insulating conventional polymer backbones are sp^3 hybridized and therefore form σ-bonds. The electronic structure consists of only σ-bands with a large electron energy band gap which renders these polymers electrically insulating and generally transparent to the visible light spectrum [18]. In contrast, conjugated molecules consist of adjoining sp^2 (or sp) hybridized carbon atoms with alternating single and multiple bonds. Each carbon atom resides in one p-orbital and three σ-bonds. The π-overlap of neighboring p-orbitals generates π-states which are delocalized along the polymer chain. The resulting relatively narrow π-band gap, which is defined by a highest occupied molecular orbital (HOMO) and lowest unoccupied molecular orbital (LUMO), is associated with low-energy electronic excitations and semiconductor behavior. The delocalized π-electron system is further responsible for the ease with which the polymer chain can be oxidized or reduced. Provided that the system is (electro)chemically doped, it exhibits high carrier mobilities that result in high electrical conductivities. However, the charge carrying species are not free electrons or holes, but quasi-particles [18].

Despite the similarity in the terminology, doping of conjugated polymers is fundamentally different from that occurring in inorganic semiconductors. The former is based on an oxidation (for p-doping) or a reduction (for n-doping) of the polymer which is accompanied by a charge-balancing incorporation of ions into the polymeric material. Doping leads to new electronic states, so-called polarons, within the otherwise forbidden electron energy gap [18]. Low doping levels give rise to singly charged polarons, while at higher concentrations, they may pair up and form spinless, doubly charged entities called bipolarons. Polarons and bipolarons were identified as the intrachain charge carriers responsible for the one-dimensional conductivity observed along one polymer chain. Additionally, they are also considered as the main contributors to interchain charge transfer between neighboring chains based on interchain hopping which is believed to be supported by the incorporated ions present in doped polymeric materials [19].

Due to their (semi)conducting properties which are highly dependent on doping and other chemical/physical factors, conjugated polymers have been applied in various devices, such as organic photovoltaic cells, electrochromic displays, chemical sensors, supercapacitors, batteries, and light emitting diodes. Here, the discussion will mainly focus on the use of DG-structured PPy in chemical sensors, since the utilization of PEDOT and PT derivatives in batteries and solar cells, respectively, is subject to ongoing research.

7.1.2 Chemical Sensors

A variety of intrinsically conducting polymers possessing sensing capacities have already been applied in several types of sensors [20]. Several transduction signals are known and are of optical, potentiometric, amperometric, or conductimetric nature. Of special interest are chemical sensors which can detect low levels of harmful or toxic gases and/or vapors. The basic sensor architecture based on electrical responses consists of a thin uniform polymer film lying on top of a pair of coplanar electrodes supported by an insulating substrate [21]. A number of theories have been proposed that aim to explain the mechanisms responsible for triggering the transient response of the active polymeric electrode material on contact or swelling with the analyte molecules. Firstly, the polaron interchain hopping and therefore the conductivity or resistivity of the polymeric material is believed to be affected by an increasing distance between neighboring chains resulting from swelling by the molecules to be detected. Secondly, swelling by the latter may also induce a change in the polymer's dielectric constant. Finally, in some cases, the analyte is expected to induce secondary doping and/or a change in morphology/cristallinity which will typically result in a conductivity or resistivity change [22–24].

Bartlett et al. presented three basic models to explain the time-dependent transient resistivity response $R(t)$ of PPy sensors to alcohol vapors [22–24]. Two of the models assume a diffusion controlled penetration of the analyte molecules into the polymer film. More specifically, they consider bounded planar as well as bounded spherical diffusion. For bounded diffusion in a plane sheet with thickness l and a diffusion coefficient D they find [23]

$$R(t) = R_\infty - (R_\infty - R_0)\sum_{n=0}^{\infty} \frac{8}{(2n+1)^2\pi^2} e^{-D(2n+1)^2\pi^2 t/4l^2}, \tag{7.1}$$

where R_0 is the initial film resistance and R_∞ is the steady-state value in presence of a solvent sensor. The corresponding expression for diffusion into a sphere of radius a is [23]

$$R(t) = R_\infty - (R_\infty - R_0)\sum_{n=0}^{\infty} \frac{6}{n^2\pi^2} e^{-Dn^2\pi^2 t/a^2}. \tag{7.2}$$

For the third model they assume sorption of the vapor molecules on free sites either within the polymer film or on its surface. Considering a fixed number of sites and negligible interactions between sites they find [23]

$$R(t) = R_\infty - (R_\infty - R_0)e^{-kt}, \tag{7.3}$$

where k is an overall kinetic constant. Besides a short response time t, the fractional change in resistance or sensitivity S is a second important figure of merit for sensors, which is given by

$$S = \frac{R_\infty - R_0}{R_0}. \tag{7.4}$$

All of these models predict that nanostructuring the conjugated polymer films should enhance the sensing performance, including a short response time and high sensitivity. Rajesh et al. presented a general review on the progress in the development of nanostructured conducting polymers/nanocomposites for sensor applications. The discussion is limited to a few publications on PPy and PEDOT sensors [25]. In 1995, Martin reported the template-assisted patterning of electrosynthesized PPy using alumina membranes [26]. Similarly, glucose biosensor based on Pt-coated PPy nanotube arrays were fabricated [27]. By coating single-walled carbon nanotubes with PPy using chemical polymerization, An et al. [28] prepared sensors with enhanced sensitivity. Most recently it was shown that PEDOT nanofibers outperform their structureless counterparts with higher sensitivity and a lower detection limit when applied in glucose biosensors [29].

7.1.3 Electropolymerization

In 1862, Letheby [30] discussed the production of a blue substance by the electrolysis of sulfate of aniline, nowadays considered as the first reported electrochemical synthesis of a conjugated polymer, in his case poly(aniline). Since then a great number of different monomers were synthesized that form conducting polymers. The synthesis method of choice for preparing these polymers involves the anodic oxidation of a suitable monomer M into a cation radical

$$\mathrm{HMH} - e^{-} \rightarrow \mathrm{HMH}^{+\bullet}, \tag{7.5}$$

where hydrogen atoms (H) in the positions of the further polymer chain bond formation are explicitly indicated [20]. Typically, the polymerization process is initiated by the dimerization of two radical species, which proceeds through the formation of a σ-bond between two sp^3 carbon atoms and the loss of two protons

$$\mathrm{HMH}^{+\bullet} + \mathrm{HMH}^{+\bullet} \rightarrow \mathrm{HMHHMH}^{2+} \rightarrow \mathrm{H(M)_2H} + 2\mathrm{H}^{+}. \tag{7.6}$$

The reaction mechanism of the proceeding electropolymerization process is still not fully understood, but the various proposed mechanisms can be summarized as

$$\mathrm{H(M)_nH}^{+\bullet} + \mathrm{H(M)_mH}^{+\bullet} - 2e^{-} \rightarrow \mathrm{H(M)_{(n+m)}H} + 2\mathrm{H}^{+}. \tag{7.7}$$

The values of n and m are subject to a controversial discussion. The classic vision of chain propagation growth assumes stepwise monomer ($n = 1$) addition to the growing chain ($m > 1$). This theory is challenged by the so-called oligomer approach which proposes that dimers ($n = 2$) or oligomers ($n > 2$) are formed as intermediates before being added to the growing polymer ($m > 2$). Two reasons make the chain propagation based on the successive coupling of monomeric cations with the chain extremely unlikely. Firstly, the rate constant of radical cation dimerization is extremely high, and secondly, the rate-determining step seems to be the elimination of protons from the intermediate σ-dimer [31]. Regardless of the underlying electrosynthesis mechanism, the complete reaction equation is given by

$$(k+2)\mathrm{HMH}^{+\bullet} - (2k+2+x)e^{-} \rightarrow \mathrm{H(M)_{(k+2)}H}^{x+} + (2k+2)\mathrm{H}^{+}. \tag{7.8}$$

As the monomer oxidation potential is always higher than the charging of oligomeric intermediates or the resulting polymer, polymerization and primary doping occur simultaneously. This electrochemical doping or oxidation requires an additional x electrons, as indicated in Eq. (7.8) [31]. In general, the resulting doping level is such that every third to fourth monomeric subunit is charged. As mentioned previously, the charging process is accompanied by a charge-balancing ion incorporation. However, several applications require the deposited polymers to be semi-conducting rather than metallically conducting and they are therefore dedoped either electrochemically or with the help of strong reducing agents, such as hydrazine.

Since the low molecular weight intermediates, such as dimers and short oligomers, are typically soluble in the electrolyte they do not immediately contribute to the weight gain at the working electrode when formed. Nevertheless, the solubility diminishes drastically with increasing chain length which finally leads to precipitation of the oligomers over a certain polymerization threshold. For most conjugated polymers this threshold in common solvents is typically quite low. After precipitation onto the electrode, possible reactions involving individual chains can lead to higher degrees of polymerization or interchain cross-linking.

The oxidation potentials of PPy and EDOT are reasonably low. Thus, the electropolymerization can be performed using less electrochemically stable solvents such as water without risking unwanted side-reactions. On the other hand, the electropolymerization of thiophene and its derivatives cannot typically be performed in pure aqueous media, since their high oxidation potential exceeds the voltages necessary for the electrolysis of water. Although the oxidation potential of thiophenes is reduced when using oligomers as monomeric precursors, such as 2,2′-bithiophene instead of thiophene, the electrosynthesis nonetheless is performed in organic media rather than water, because of low monomer solubility in the latter, and because of the fact that the cation radical is extremely reactive towards nucleophilic species [32, 33]. This combined with the fact that the styrenic polymer template is dissolved by most of the organic solvents commonly used for the electrosynthesis of PT, including acetonitrile and boron trifluoride diethyl etherate, complicates the search for an appropriate electrolyte solvent [34].

7.1.4 Polymer Solubility and Solubility Parameters

Unlike non-polymeric materials, polymers do not dissolve instantaneously due to chain entanglement and dissolution proceeds in two steps, solvent diffusion into the polymeric material and subsequently chain disentanglement [35]. Furthermore, certain solvents tend to swell particular polymers but do not posses the necessary solubility strength to achieve complete chain disentanglement and dissolution. Swelling of the polymeric material by the molecules of these poor or ‘non-solvents’ often induces osmotic stress cracking or plasticization. The latter generally leads to structural changes on the nanoscale due to an increased plasticity which is generally accompanied by a decrease of the glass transition temperature [36].

In many industrial applications polymer processing plays a key role and therefore theories predicting the interaction of solvent molecules with polymer chains have been developed that aid in the selection of suitable solvents.

As in the case of polymer-polymer blends, the dissolution of a polymer in a solvent is governed by the free energy of mixing, see Eq. (4.1). The positive entropic contribution associated with the dissolution of a high molecular weight polymer in a solvent is higher than the one arising from the formation of a homogenous polymer melt. Nevertheless, this contribution is still very small compared to the enthalpy term that therefore determines the sign of the free energy change. Various models based on solubility parameters describing the interaction of polymers with solvents were developed to provide an estimate for the enthalpy of mixing. To date, Hansen's model delivers the most accurate predictions concerning solubility using three solubility parameters δ_{D}, δ_{P}, and δ_{H}, which describe the energy contributions from dispersion forces between induced dipoles, from interaction forces between permanent dipoles, and from hydrogen-bonding between molecules, respectively [37]. Hansen further defined the following norm that allows predictions concerning the miscibility of two substances A and B

$$R_{\mathrm{AB}} = \sqrt{4(\delta_{\mathrm{D,A}} - \delta_{\mathrm{D,B}})^2 + (\delta_{\mathrm{P,A}} - \delta_{\mathrm{P,B}})^2 + (\delta_{\mathrm{H,A}} - \delta_{\mathrm{H,B}})^2}. \tag{7.9}$$

If R_{AB} lies below a certain threshold R_0, the substances are predicted to be miscible, i. e. the polymer is predicted to be soluble in the solvent. On the other hand, the solvent will neither dissolve nor swell the solute in case of $R_{\mathrm{AB}} \gg R_0$. Furthermore, the Hansen solubility parameters can be used for estimating the Flory-Huggins interaction parameter of two polymers

$$\chi = \alpha \frac{R_{\mathrm{AB}}^2}{RT}, \tag{7.10}$$

where $\alpha \approx 4$ is a scaling constant, R the gas constant, and T the temperature [38]. The Hansen solubility parameters for selected polymers and solvents are listed in Table 7.1 (Table 7.2), while the calculated R_{AB} are given in Table 7.3.

7.2 Experimental Methods

7.2.1 Materials

The chemicals and materials listed in Table 7.2 were used as received unless stated otherwise.

7.2.2 Electropolymerization

Voided templates prepared on FTO-coated glass using either P(F)S49-*b*-PLA192 $\Big|^{174\,^{\circ}\mathrm{C}}_{0.33\,\mathrm{h}}$ or PS50-*b*-PLA214 $\Big|^{174\,^{\circ}\mathrm{C}}_{0.33\,\mathrm{h}}$ were successfully employed in the template-assisted nanopatterning of conjugated polymers.

Table 7.1 Experimentally determined and theoretically calculated Hansen solubility parameters for selected polymers and solvents

Solvent/polymer	δ_D [$MPa^{0.5}$]	δ_P [$MPa^{0.5}$]	δ_H [$MPa^{0.5}$]	R_0 [$MPa^{0.5}$]	Reference
PS	21.3	5.75	4.3	12.7	[37]
PFS	19.4	3.4	3.6	–	[a]
PPy	18.9	12.9	10.4	–	[a]
PT	20.2	4.7	9.6	–	[a]
P3MT	18.9	3.3	11.2	–	[a]
P3HT	17.9	3.2	0.91	3.3	[39]
P3HT	17.4	1.1	0.5	–	[a]
PEDOT	21.2	7.4	8.1	–	[a]
PCBM	19.3	7.3	8.02	6.8	[39]
Acetone	15.5	10.4	7.0	–	[37]
Acetonitrile	15.3	13.0	6.1	–	[37]
Benzene	18.4	0.0	2.0	–	[37]
Cyclohexane	16.8	0.0	0.2	–	[37]
Diethyl ether	14.5	2.9	5.1	–	[37]
Ethanol	15.8	8.8	19.4	–	[37]
Ethanolamine	17.2	15.6	21.3	–	[37]
Ethylene carbonate	19.4	21.7	5.1	–	[37]
Formamide	17.2	26.2	19.0	–	[37]
Hexane	14.9	0.0	0.0	–	[37]
Isopropanol	15.8	6.1	16.4	–	[37]
Methanol	15.1	12.3	22.3	–	[37]
Toluene	18.0	1.4	2.0	–	[37]
Water	15.6	16.0	42.3	–	[37]
Xylene	17.8	1.0	3.1	–	[37]
Chlorobenzene	19.0	4.3	2.0	–	[37]
[BMIM][BF_4]	23.0	19.0	10.0	–	[37]

[a] Estimate calculated using the HSPiP software

7.2.2.1 Poly(bithiophene) and Poly(3-methylthiophene)

The electropolymerization of thiophene-derivatives was conducted in a glove box using a two-electrode setup equipped with a Cu-mesh as counter electrode. The electrolyte used was the ionic liquid [BMIM][TFSI], which was deoxygenated by bubbling with nitrogen for 40 min and dried overnight with molecular sieves prior to use, containing 0.2 M of either 2,2′-bithiophene or 3-methylthiophene [40]. Electrosynthesis was conducted under potentiostatic conditions at 3.2–3.4 V and 3.6–3.7 V for bi- and methylthiophene, respectively. Thereafter, the films were successively rinsed in [EMIM][BF_4] and DI water, dried, and freed from their templated by dissolution in pure diethyl ether or a 2:1 mixture of diethyl ether and hexane.

Table 7.2 Chemicals and materials used in this part of the study

Chemical	Purity [%][a]	Abbreviation
1-Butyl-3-methylimidazolium bis(trifluoromethylsulfonyl)imide	98 (SA)	[BMIM][TFSI]
1-Ethyl-3-methylimidazolium tetrafluoroborate	98 (SA)	[EMIM][BF_4]
2,2′-Bithiophene	97 (SA)	–
3,4-Ethylenedioxythiophene	97 (SA)	EDOT
3-Methylthiophene	98 (SA)	–
Diethyl ether	– (SA)	–
Hexane	– (FS)	–
Hydrazine hydrate	50–60 (SA)	–
Lithium perchlorate	95 (SA)	–
Methanol, laboratory reagent grade	– (FS)	–
p-Toluenesulfonic acid monohydrate	98 (SA)	–
Pyrrole	98 (SA)	–
Sodium chloride	99.5 (SA)	–
Thiophene	99 (SA)	–
Toluene, anhydrous	99.8 (SA)	–
Water, deionized 18 MΩ	–	DI water
Xylene	– (FS)	–

[a] Suppliers are given in brackets: Fisher Scientific, FS; Sigma Aldrich, SA

Table 7.3 Calculated values of R_{AB} for selected polymers and solvents

	PS	PPy	PT	P3MT	PEDOT	PCBM
Acetone	13 •	8	11	11	12	8 •
Acetonitrile	14 •	8	13	13	13	10
Benzene	8 •	15	10	10	11	10
Cyclohexane	11 ⋆	17	13	12 ⋆	14	12 ○
Diethyl ether	14 •	14	12 ⋆	11 ⋆	14 ⋆	11
Ethanol	19 ⋆	12	14 ⋆	12 ⋆	16 ⋆	13
Ethanolamine	21 ⋆	12	17	16	17	16
Ethylene carbonate	16 ○	10	18	19	15	15
Formamide	26 ⋆	16	24	24	23	22
Hexane	15 ⋆	18	15 ⋆	14 ⋆	17	14
Isopropanol	16 ⋆	11	11	9	14	11
Methanol	23 ⋆	14	18 ⋆	16 ⋆	19 ⋆	17
Toluene	8 •	14 ○	9 ○	10 ○	11 ○	9
Water	41 ⋆	33 ⋆	36 ⋆	34 ⋆	37 ⋆	36
Xylene	9 •	14 ⋆	9	9	11 ○	9
Chlorobenzene	5 •	12	8 ○	9 ○	8	7 ⋆
[BMIM][BF_4]	15 ⋆	10	15	18	12	14
[BMIM][TFSI]	⋆		⋆	⋆		

⋆ Good solvent, dissolves the polymer film completely. ○ Poor solvent, does not fully dissolve the polymer film, but destroys the nanostructure. • Non-solvent, does not seem to swell the polymer film at all, since the polymer nanostructure stays intact

7.2.2.2 Poly(3,4-ethylenedioxythiophene)

PEDOT was electrosynthesized according to Du and Wang using an aqueous solution containing 0.02 M EDOT and 0.1 M $LiClO_4$ [41]. The monomer was dissolved with the help of an ultrasonic bath. A standard three-electrode setup with a Pt-mesh counter electrode and an Ag/AgCl reference electrode was employed. The styrenic templates were rinsed with methanol before immersing the wet samples into the aqueous electrolyte. Electropolymerization was conducted under potentiostatic condition at 0.9 V. The vividly blue samples were rinsed with DI water and dried with a nitrogen stream. Diethyl ether was found to be a suitable solvent for the selective removal of the template.

7.2.2.3 Poly(pyrrole)

The same three-electrode electrochemical cell as in the case of PEDOT was used for the poly(pyrrole) electrosynthesis using an aqueous solution containing 0.1 M p-toluenesulfonic acid monohydrate and 0.1 M sodium chloride [42]. Unlike EDOT, pyrrole spontaneously oligomerizes in the presence of oxygen. Hence, the electrolyte was deoxygenated by bubbling with nitrogen prior to the addition of 0.5 M pyrrole and kept under a protective atmosphere during polymerization. Typically the polymerization was performed at room temperature, but some films were also synthesized at a low temperature (5 – 10 °C). The hydrophobic templates were once again rinsed with methanol before placing them opposite the counter electrode into the deposition bath. The electropolymerization was performed potentiostatically at 0.65 V and the deposit thickness was controlled by monitoring the charged passing through the cell. Before selectively dissolving the redundant styrenic template with xylene the samples were rinsed with DI water and dried.

7.2.3 Preparation of Gas Sensors Based on Poly(pyrrole) Films

The resistance gas sensor design consisted of two coplanar chromium contacts which were evaporated onto an insulating glass substrate and a PPy film was placed on top of them, bridging the gap, see Fig. 7.2. This architecture allows to monitor the resistivity change of the sensing material, which connects the two separated electrodes, as a response to gases or solvent vapors.

A first batch of sensors was assembled in the following way. First an as-deposited PPy film prepared on one of the usually used FTO-coated substrates was coated with a thick layer of high molecular weight poly(styrene) dissolved in xylene. After drying the sample on a hotplate at 50°C the combined film was carefully peeling off and transferred onto a chromium-patterned substrates. Finally the poly(styrene) was dissolved by washing in xylene for several hours.

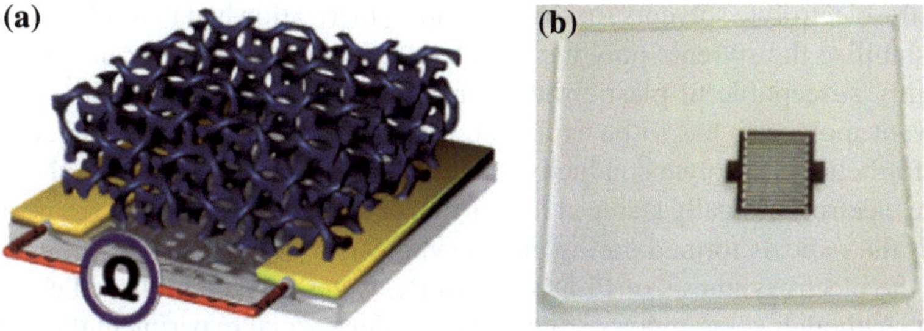

Fig. 7.2 **a** Double-gyroid gas sensor design consisting of two coplanar metal electrodes on an insulating substrate. **b** Interdigitated metal electrodes evaporated onto a microscope slide

A second, more sophisticated type of devices was fabricated in the following way. First a temporary gold layer, which closed the nonconducting gap between the two chromium electrodes, was evaporated. Subsequently a voided double-gyroid-structured template was prepared on this continuously metal-coated substrate and refilled with PPy. Then the redundant template was dissolved with xylene. Finally the superficial gold layer was etched with a wet gold etchant, which reestablished the gap between the chromium electrodes.

7.2.4 Gas Sensors Testing Rig

For the testing of the gas sensors a setup built by Dr. Richard White and Dr. Sven Hüttner was used, which consisted of a solvent bubbling unit, a potentiostat-galvanostat for recording the transient resistance response, and a vacuum chamber which fitted with a special sensors chip holder. The resistance response was determined by applying a fixed potential across the two evaporated electrodes, typically 0.1 V, and measuring the time dependent electric current while introducing solvent vapors into the chamber. The solvent bubbling unit was built with two digital mass flow controllers regulating the influx of dry nitrogen and solvent vapor-saturated nitrogen into the chamber and a temperature bath to adjust the solvent's vapor pressure.

7.3 Results and Discussion

7.3.1 Identifying Suitable Solvents

As mentioned above, suitable orthogonal solvents had to be identified for the different, successive fabrication steps of functional devices based on nanostructured conjugated polymer films.

Firstly, electrolyte solvents for the electropolymerization had to be chosen that do not destabilize the styrenic polymer templates. Note that polymeric nanostructures are highly susceptible to plasticization and the slightest swelling of the template by solvent molecules has to be excluded. Simultaneously, the monomers and salts added for better electrolyte conductivity have to be soluble. Importantly, the solvent must be electrochemically stable at the particular deposition potential and be neutral towards the radicals formed during electrosynthesis.

For the electrosynthesis of PEDOT and PPy, water was the preferred electrolyte solvent. Although some reports suggest that under special experimental conditions the aqueous electropolymerization of poly(thiophene) derivatives is possible, all attempts conducted in this study failed, including deposition at low pH values and using bithiophene which exhibits a lower oxidation potential than thiophene as the monomeric species [43, 44]. Typically, boron trifluoride diethyl etherate and (fairly) anhydrous acetonitrile were used instead [34]. Although both solvents were successfully used to prepare poly(thiophene) inverse opals via the templated synthesis using poly(styrene) microsphere arrays, they tended to destroy the styrenic DG-structured scaffold [2–4].

The double-gyroid templates were found to be stable in a number of polar protic solvents, but these do not support the electrosynthesis of poly(thiophene), since the intermediate radical is extremely reactive towards nucleophilic species. Further, non-polar solvents such as hexane and various fluorinated solvents were compatible with the styrenic templates, but unfortunately common ionic salts, which are usually added to increase the electrical conductivity of the electrolyte are not soluble in these.

Finally, room temperature ionic liquids were identified to be suitable. Especially the commercially available [EMIM][BF_4] and [BMIM][TFSI] seemed adequate due to their relatively low viscosity at room temperature. Advantageously, [BMIM][TFSI] wetted the hydrophobic mesopores of the styrenic template, making the addition of surfactants to lower the surface tension redundant. On the other hand, [EMIM][BF_4] did not infiltrate the pores readily, but dipping the templates into [BMIM][TFSI] prior to immersion into [EMIM][BF_4] circumvented this issue. In many applications the extremely low vapor pressure of ionic liquids is advantageous, but in this particular case it makes the removal of the electrolyte after electrosynthesis more sophisticated. Especially, because [BMIM][TFSI], which was preferentially used for electropolymerization is neither soluble in water nor the solvents used for template removal. Primary alcohols were found to be miscible with [BMIM][TFSI], but the conjugated polymer films tended to crack on immersion into the alcohols. This is most probably osmotic stress-cracking due to swelling of the conjugated polymer by the primary alcohol molecules, extent of which could be reduced by using alcohols with larger alkyl groups. A solution to this problem was to dip the electropolymerized samples into [EMIM][BF_4], which is miscible with both [BMIM][TFSI] and water. The films were then rinsed in the latter to remove the water soluble ionic liquid and subsequently dried on a hotplate. It is worth mentioning, that in some cases the films cracked during washing in water. Curiously, immersion of the samples into [EMIM][BF_4] led to a color change, most likely induced by an ion exchange or (un)doping of the conjugated polymer.

Secondly, appropriate orthogonal solvents for the selective template dissolution had to be found. Because of the same reason mentioned above, these solvents are not allow to even slightly swell the nanostructured conjugated polymers, otherwise the desired structural features are inevitable lost. Diethyl ether and xylene were identified as ideal solvents for PEDOT and PPy, respectively. Chlorobenzene was reported to have no effect on the nanostructure of electrodeposited P3MT, but in this study the opposite was found [13]. Instead, pure diethyl ether or a 2:1 mixture of diethyl ether and hexane were used.

Lastly, solvents for dedoping and refilling of the mesoporous conjugated polymer films with a second soluble functional material are of interest. For example, undoped, semi-conducting poly(thiophene) derivatives can be applied in bulk heterojunction solar cells in combination with n-type PCBM. Cyclohexane was found to partially dissolve PCBM while only swelling PT and P3MT nanostructures. According to the Hansen solubility parameters acetone and nitromethane are other promising candidates, but they were not yet tested.

With one exception, the solubility predictions based on the Hansen solubility parameters were fairly consistent with the experimental findings of this study. The results together with the calculated R_{AB} values for several solvents are listed in Table 7.2. Note that the Hansen sphere radius $R_0 = 12.7$ of poly(styrene) is expected to be considerably larger than that of conjugated polymers.

7.3.2 DG-Structured Poly(bithiophene) and Poly(3-methylthiophene)

As mentioned above, all attempts of electropolymerizing thiophene, bithiophene, and methylthiophene from aqueous solutions failed in this study. Performing the synthesis in an anhydrous and oxygen-free environment using [BMIM][TFSI] as electrolyte, finally gave the desired results. Sometimes it was necessary to apply an overpotential of a few hundred millivolts for about 1 s in order to nucleate and initiate the polymerization. No obvious morphological differences between DG-structured PT and P3MT films were observable when imaging with an electron microscope.

Figure 7.3a shows the profound impact a so-called 'non-solvent' for PT, such as chlorobenzene, can have on the nanostructure. Although chlorobenzene does not dissolve the polymer film, no obvious mesoporous nanostructure is left after template removal. This is most probably due to diffusion of solvent molecules into the conjugated polymer which destabilize the about 11 nm thick gyroidal network struts and eventually leads to the structural collapse. Using diethyl ether as solvent for the template dissolution instead resulted in mesoporous and highly ordered films. However, the free-surface did not show the same degree of order and porosity as the bulk of the film, see Fig. 7.3c. This behavior at the free-surface was more pronounced in the case of PEDOT and PPy, and will be discussed in more detail in the following to sections.

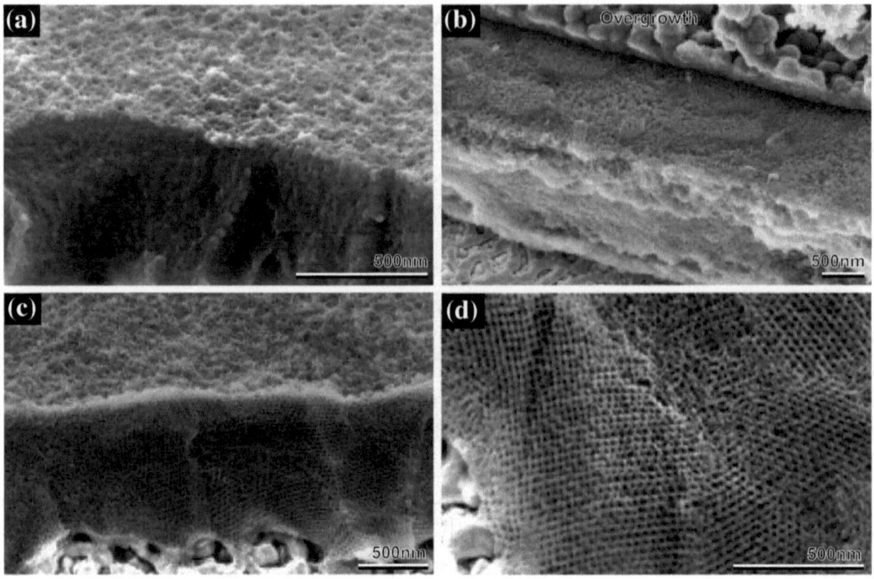

Fig. 7.3 SEM images of a DG-structured PBT after template removal. **a** Chlorobenzene used for template dissolution results in a loss of the structure, leaving only a few recognizable pores. Diethyl ether or a hexane/diethyl ether mixture were used as selective template solvents for the remaining SEMs. **b** Cross-sectional view onto a film prepared by overfilling a PS50-*b*-PLA214$\big|^{174\,°C}_{0.33\,h}$ template. The template's semi-porous free-surface allowed the partial removal of the overgrowth before template dissolution resulting in a smooth free-surface. **c** Cross-section revealing a well-defined DG nanomorphology in the bulk, while the free-surface looks more disordered. **d** Magnification of a highly ordered cross-section

As discussed previously, when electropolymerizing conjugated polymers they are inevitably doped during the synthesis process by incorporation of $[TFSI]^-$ ions. Therefore, dedoping the films becomes necessary before potential luminescent or photovoltaic applications can be realized. Chemical dedoping with hydrazine as reducing agent and electrochemical dedoping were tested. Altering the polymer's oxidation state in this way is typically accompanied by a chromogenic response, see Sect. 5.1.3. As-deposited PT films are brown/black and P3MT deposits are a dark blue, while both are brown/red in their dedoped state. Measuring the optical absorbance or transmittance spectra of a polymer film before and after (de)doping is a noninvasive way to obtain valuable information about the doping state, see Fig. 7.4. The mesoporous nature of the double-gyroid-structured conjugated polymer allows to effectively dedope micrometer thick films, since the incorporated anions are located within 6 nm of the polymer/solvent interface. Importantly, the mesoporous gyroidal nanostructure was still preserved after dedoping, as confirmed by SEM, regardless which technique was used. Using these undoped films in conjunction with PCBM in photovoltaic devices is subject to ongoing work.

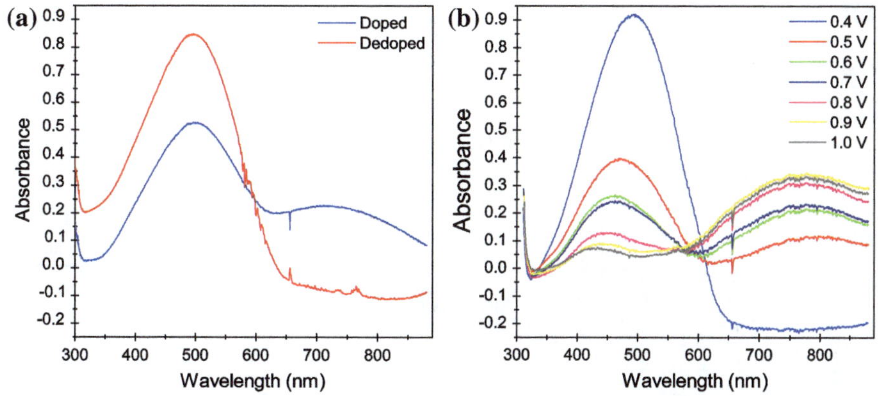

Fig. 7.4 Absorption spectra of doped and dedoped DG-structured P3MT films on FTO-coated glass. Negative absorbance values are most likely due to a higher reflectance of the bare FTO substrate used for calibration compared to the dull polymer films. **a** Chemical dedoping for 1 min in an aqueous hydrazine solution (50–60 %). Since the films are handled in air after reduction they might be subject to oxidation before recording the 'dedoped' spectrum. **b** Electrochemical dedoping performed in an aqueous 0.1 M $LiClO_4$ solution. Compared to (**a**) these spectra were measured in situ. The data was recorded by Harry Benison and Michael Price

7.3.3 DG-Structured Poly(3,4-ethylenedioxythiophene)

The fabrication of double-gyroid-structured blue PEDOT deposits on FTO-coated glass was highly reproducible. As discussed in Sect. 7.1.3 the dimers and oligomers which are intermediately formed during electrosynthesis only become insoluble and precipitate once they reached a certain degree of polymerization. Further, oligomeric radicals may not react with a second of their species because of a low monomer concentration in close vicinity of the electrode due to diffusion limitations through the mesoporous template. This gives us reason to believe that oligomeric species may diffuse away from the electrode and then absorb to the walls of the narrow styrenic templated. This in turn means that the conjugated polymer growth proceeds successively, initially an intermediate nanotubular deposit is formed by coating the voided template channels, before the latter are fully filled and finally a network of nanowires is established. Since the nanotubular growth front does not show the same stability as the network of solid nanowires, they tend to collapse on template dissolution and thus, agglomerate as a semiporous layer covering the polymer film surface, see Fig. 7.5.

For certain applications a highly porous free-surface might be of importance, thus the possibility to etch the semiporous layer using a plasma etcher (Diener MRC 100) was tested. Figure 7.5b, c show the successful results obtained after a 10 and 20 s air plasma treatment ignited at a pressure of 0.4 mbar and with the generator power set to 20 %.

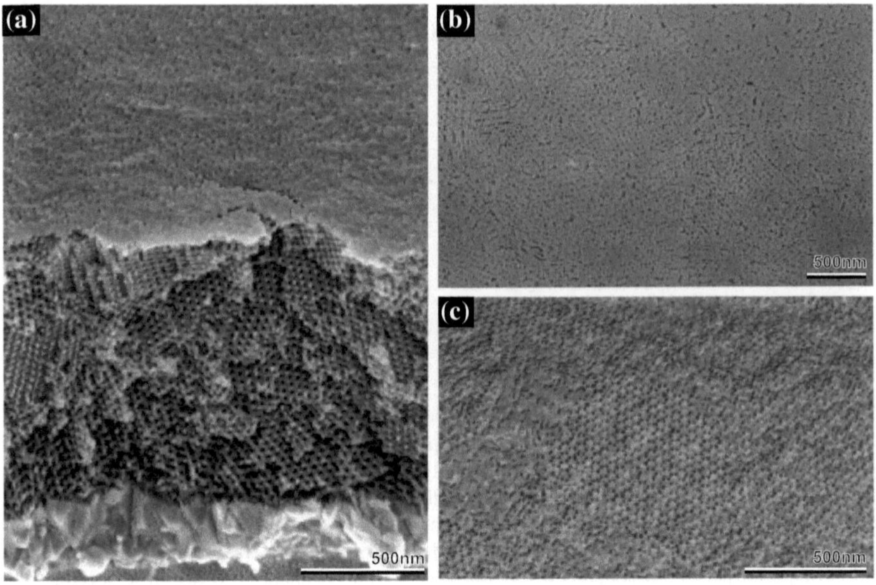

Fig. 7.5 SEM images of DG-structured PEDOT after selective template removal with diethyl ether. **a** Cross-section showing a highly porous and well-ordered structure in the bulk of the film. However, the free-surface does not show the same high degree of porosity. **b, c** Free-surfaces after a 10 and 20 s air plasma treatment, respectively, resulting in an improved interfacial porosity

7.3.4 DG-Structured Poly(pyrrole)

Similarly successful was the preparation of pitch black double-gyroid-structured PPy films from aqueous solutions. Typically a current density of about 1 mA cm^{-2} was observed at an applied potential of 0.65 V vs. Ag/AgCl. Recorded CVs were consistent with the ones presented in the literature [45]. Additionally, for comparison, nontemplated films which also showed some structural features were prepared for the further use as active material in gas sensors, see Fig. 7.6. The relationship between the total charge density consumed during electropolymerization and the thickness of the film, which was measured by SEM imaging, was investigated. A total charge density of 1 C cm^{-2} was found to give rise to an average thickness of about 4 and 10 μm for nontemplated and double-gyroid-structured films, respectively.

Similarly to PEDOT, thick PPy films (1.5 μm) tended to form an about 10 nm thick semiporous layer covering the free film surface, while thin films (350 nm) did not exhibit this unwanted feature, compare Figs. 7.7 and 7.8. The same plasma etching conditions as used in Sect. 7.3.3 proved to be effective to remove this layer. The etch rate determined for nontemplated PPy films under these conditions was on the order of 0.1 μm min^{-1}.

The sheet resistance of both nontemplated and double-gyroid-structured films was determined by measuring the current through a 1 cm long film section for a

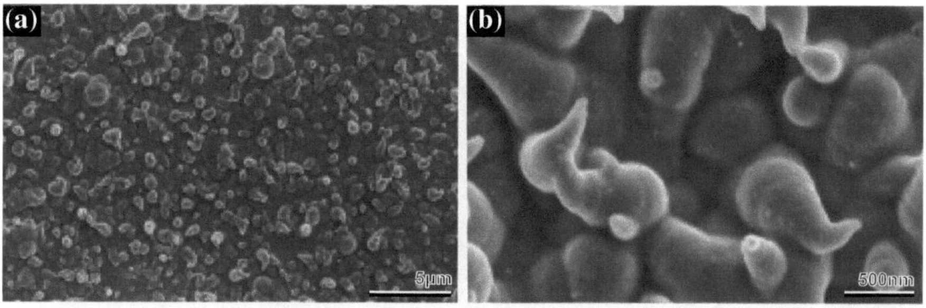

Fig. 7.6 **a** SEM of the free-surface of a nontemplated PPy film exhibiting some structural features. **b** Magnified view of (**a**)

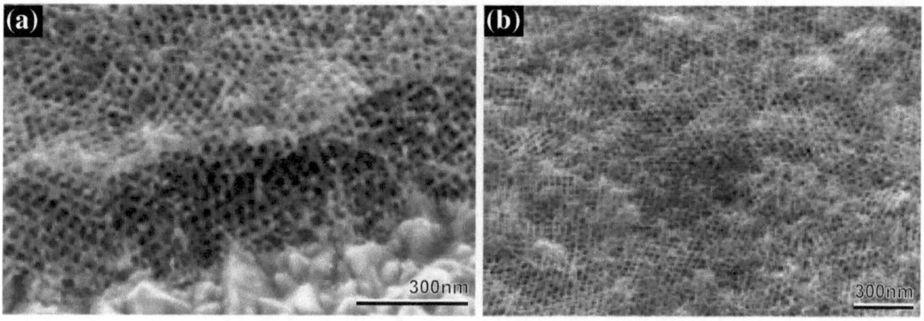

Fig. 7.7 SEM images of a 350 nm thin DG-structured PPy film after selective template removal with xylene. **a** Cross-sectional view of a thin film with the native free-surface being highly porous. **b** Top-view onto the corresponding free-surface (no plasma treatment)

known applied potential difference, omitting any geometric factors. Moreover, possible contact resistance arising from the contact points between the film and the probes was neglected. The measured values are therefore only an upper estimate for the resistance. Regardless of the film structure, freshly prepared samples exhibited a sheet resistance on the order of $50 - 100\,\Omega/\Box$. This translates to a low electrical resistivity of $0.01\,\Omega\,\text{cm}$. When stored in ambient air the resistance of the films increased with time.

7.3.5 Testing the Chemical Sensors Based on Poly(pyrrole)

Both gas sensors based on nontemplated and nanostructured PPy films were found to respond to a variety of solvent vapors, including methanol, ethanol, isopropanol, acetone, and also to humidity with a resistance increase. Initial tests were performed in ambient air leading to the observation of the following phenomenon, the sensors responded to water vapor with an immediate resistance increase, while exposure to organic solvents led to an initial small decrease in resistance before showing the

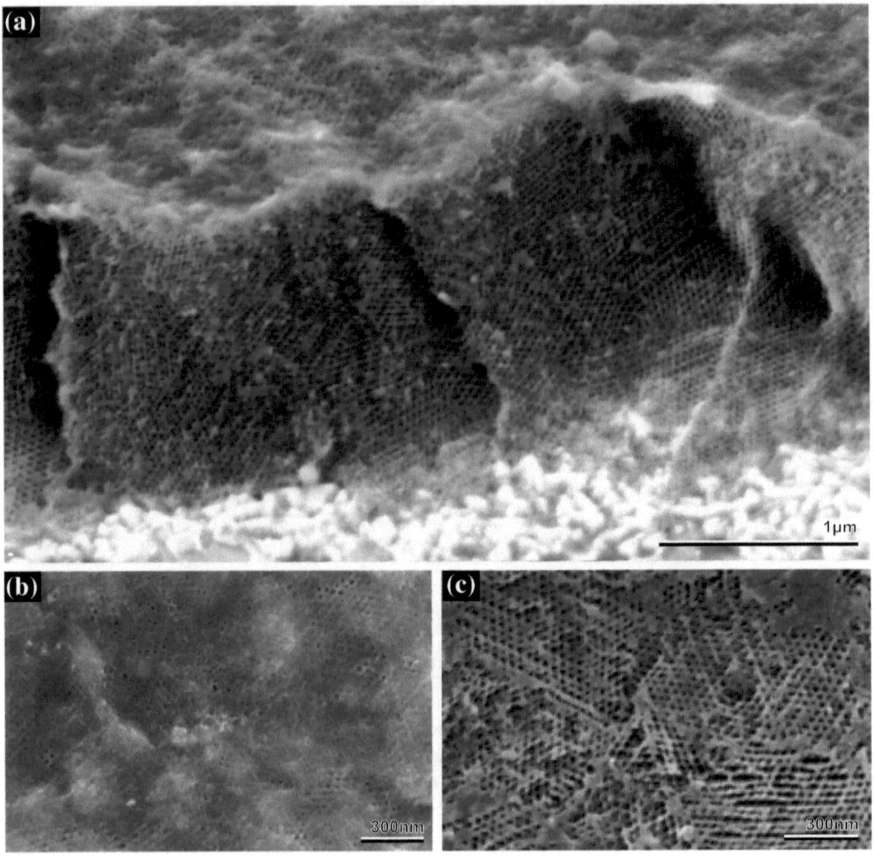

Fig. 7.8 SEM images of a 1.5 μm thick DG-structured PPy deposit after selective template removal with xylene. **a** Cross-section revealing a highly porous and well-ordered structure in the bulk of the film and a semiporous free-surface. **b** Top-view onto the semiporous free-surface. **c** Corresponding free-surface after a 20 s air plasma treatment

expected increase. When repeating the same experiments in a dry environment this abnormality did not occur. These findings suggest, that water and analyte molecules are competing for the available sensing sites and hence, all following measurements were conducted under a dry nitrogen atmosphere.

PPy films were allowed to equilibrated for a few minutes under vacuum or in dry nitrogen until a stable baseline resistance, typically between 0.1 and 1 kΩ, was reached before starting the measurement. This initial low 'baseline' resistance was typically not restored after ending the first solvent exposure and nitrogen purge cycles, regardless of the extent of the purge. The subsequent exposures, however, were fairly reversible and reproducible, although some samples displayed a drift over time. Figure 7.9 shows the time-dependent transient response of nontemplated and DG-structured films on alternating exposure to alcohol vapors, such as methanol

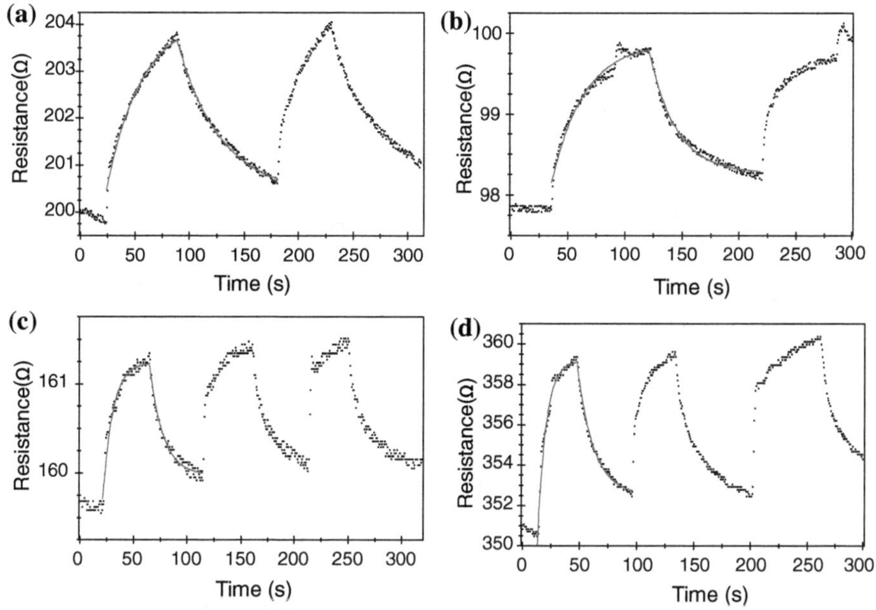

Fig. 7.9 Resistance response of nontemplated and DG-structured PPy gas sensors to different alcohol vapors and purging with dry nitrogen. Nitrogen was used as carrier gas. Equation 7.3, describing the sorption model, was used to calculate the *blue* fitting curves. **a** Nontemplated film, 11 % methanol vapor. **b** Nontemplated film, 4 % ethanol vapor. **c** DG-structured film, 1 % ethanol vapor. **d** DG-structured film, 4 % ethanol vapor

and ethanol, and purging with nitrogen. The data was fitted using Eqs. (7.1–7.3), although the sorption model was found to give the best fit, the diffusion models also produced acceptable fitting curves. Our PPy films showed similar relative response times to the different alcohols, contrary to Bartlett et al. who measured significantly longer response times for the higher molecular weight alcohols [22–24].

The response speed, associated with the adsorption of vapor molecules on free sites in the PPy films, was typically found to be faster than the recovery speed, consisting of the desorption and evaporation of the analyte molecules from the occupied sensing sites, see Fig. 7.10a. This became especially evident in nontemplated films. On average the response time of nontemplated and DG-structured films on exposure to solvent vapors was similar, but on purging with nitrogen the structured films outperformed the nontemplated deposits, see Fig. 7.10a. Curiously, the response times varied very little with different vapor concentrations.

Figure 7.10b illustrates the results of sensitivity measurements for the resistance sensor exposed to various concentrations of ethanol, see Eq. 7.4. Both nontemplated and nanostructured films seem to exhibit a 'linear' relationship between the analyte's partial pressure and the measured sensitivity [46]. The best performing devices reached a sensitivity of only 2 and 2.5 % for a nontemplated and a DG-structured film, respectively, when tested in a nitrogen atmosphere containing

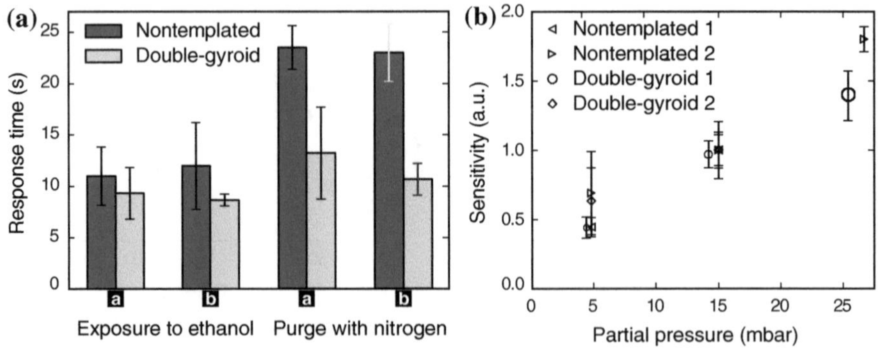

Fig. 7.10 **a** Response time needed to complete a change of $\Delta R = (R_\infty - R_0)/2$ obtained from fitting the transient curves with Eq. 7.3. The nontemplated and DG-structured film were electropolymerized under the same conditions including the same total current density. Data sets a and b correspond to exposures to 1.2 and 3.8 % of ethanol vapor in dry nitrogen, respectively. **b** Sensitivity of nontemplated and DG-structured film as function of the partial ethanol pressure. Alexandre Nicolas analyzed the data and prepared the present graphs [7]

4 % ethanol. Contrary to our expectations, films exhibiting a mesoporous nanostructure did not outperform their nontemplated counterparts regarding sensitivity or response time. The only reasonable explanations we can provide for the disappointing performance of our gas sensors are, firstly, that the employed films of micrometer thickness were too thick, and secondly, that the doping level and therefore the conductivity of the polymer films was too high. This hypothesis is supported by the findings of Bartlett et al. who reported that the sensitivity decreases with an increasing PPy film thickness [23]. Due to time constraints this was not further investigated.

7.4 Concluding Remarks

In summary, we established a fabrication route to three-dimensional conjugated polymer nanostructures. Importantly, organic scaffolds are used as templates that can be removed by dissolution with organic solvents which typically do not chemically alter the conjugated polymers. In comparison, the selective removal of inorganic templates made from silica or alumina generally require corrosive acids or strong bases that might damage the molecular structure of the polymeric material. Thereby, the selection of appropriate orthogonal solvents for the several production steps involved played a key role. The conducting polymers successfully nanostructured by this template-directed electropolymerization included PPy, PEDOT, PBT, and P3MT.

In a first study, DG-structured PPy films were tested as active sensing material in gas sensors. Although the prepared sensors were efficient in detecting solvent vapor concentrations as low as 1 % with great reliability, and with a response time below 20 s, we could not show that the performances of these was greatly enhanced by the well-ordered and highly porous nanostructure when compared to their non-templated counterparts. More specifically, neither the sensitivity nor the response time were significantly improved. Nevertheless, we are still convinced that the high interfacial surface area of the prepared mesoporous conjugated polymers will lead to a considerable enhancement of the device performance in other applications.

In this spirit, we will study the performance of structured PEDOT films in batteries and/or supercapacitors as well as the application of dedoped PT and P3MT in bulk heterojunction solar cells in the near future [12–15]. Furthermore, the nanostructured conjugated polymer films may find application in thermoelectric devices [47–50].

References

1. V.M. Cepak, C.R. Martin, Chem. Mater. **11**(5), 1363–1367 (1999)
2. P.N. Bartlett, P.R. Birkin, M.A. Ghanem, C.-S. Toh, J. Mater. Chem. **11**(3), 849–853 (2001)
3. T. Cassagneau, F. Caruso, Adv. Mater. **14**(1), 34–38 (2002)
4. T. Cassagneau, F. Caruso, Adv. Mater. **14**(24), 1837–1841 (2002)
5. J.I. Lee, S.H. Cho, S.-M. Park, J.K. Kim, J.K. Kim, J.-W. Yu, Y.C. Kim, T.P. Russell, Nano Lett. **8**(8), 2315–2320 (2008)
6. S.I. Cho, S.B. Lee, Acc. Chem. Res. **41**(6), 699–707 (2008)
7. A. Nicolas, *Controlled Assembly and Structuration on the Nanoscale* (University of Cambridge, 2011)
8. W. Wiedemann, L. Sims, A. Abdellah, A. Exner, R. Meier, K.P. Musselman, J.L. MacManus-Driscoll, P. Müller-Buschbaum, G. Scarpa, P. Lugli, L. Schmidt-Mende, Appl. Phys. Lett. **96**(26), 263109–263109-3 (2010)
9. J. Weickert, R.B. Dunbar, H.C. Hesse, W. Wiedemann, L. Schmidt-Mende, Adv. Mater. **23**(16), 1810–1828 (2011)
10. G.A. Snook, P. Kao, A.S. Best, J. Power Sources **196**(1), 1–12 (2011)
11. M.D. Ingram, H. Staesche, K.S. Ryder, Solid State Ionics **169**(1–4), 51–57 (2004)
12. E.L. Ratcliff, J.L. Jenkins, K. Nebesny, N.R. Armstrong, Chem. Mater. **20**(18), 5796–5806 (2008)
13. H. Gao, Y. Chen, Y. Luo, Nanotechnology **22**(28), 285203 (2011)
14. R. Valaski, C. Canestraro, L. Micaroni, R. Mello, L. Roman, Sol. Energy Mater. Sol. Cells **91**(8), 684–688 (2007)
15. G. Tsekouras, C.O. Too, G.G. Wallace, Electrochim.Acta **50**(16–17), 3224–3230 (2005)
16. H. Shirakawa, E.J. Louis, A.G. MacDiarmid, C.K. Chiang, A.J. Heeger, J. Chem. Soc. Chem. Commun. **16**, 578 (1977)
17. B. Bolto, R. McNeill, D. Weiss, Aust. J. Chem. **16**(6), 1090–1103 (1963)
18. W. Salaneck, R. Friend, J. Brédas, Phys. Rep. **319**(6), 231–251 (1999)
19. L. Zuppiroli, M.N. Bussac, S. Paschen, O. Chauvet, L. Forro, Phys. Rev. B **50**(8), 5196–5203 (1994)
20. S. Cosnier, *Electropolymerization: concepts, materials and applications* (Wiley-VCH, Weinheim, 2010)
21. L. Al-Mashat, H.D. Tran, W. Wlodarski, R.B. Kaner, K. Kalantar-Zadeh, IEEE Sens. J. **8**(4), 365–370 (2008)
22. P.N. Bartlett, P.B. Archer, S.K. Ling-Chung, Sens. Actuators **19**(2), 125–140 (1989)
23. P.N. Bartlett, S.K. Ling-Chung, Sens. Actuators **19**(2), 141–150 (1989)
24. P.N. Bartlett, S.K. Ling-Chung, Sens. Actuators **20**(3), 287–292 (1989)

25. T. Ahuja Rajesh, D. Kumar, Sens. Actuators B **136**(1), 275–286 (2009)
26. C.R. Martin, Acc. Chem. Res. **28**(2), 61–68 (1995)
27. E.M.I.M. Ekanayake, D.M.G. Preethichandra, K. Kaneto, Biosens. Bioelectron. **23**(1), 107–113 (2007)
28. K.H. An, S.Y. Jeong, H.R. Hwang, Y.H. Lee, Adv. Mater. **16**(12), 1005–1009 (2004)
29. K. Layton, M. Abidian, in *5th International IEEE/EMBS Conference on Neural Engineering (NER)* (2011), pp. 298–301
30. H. Letheby, J. Chem. Soc. **15**, 161 (1862)
31. J. Heinze, B.A. Frontana-Uribe, S. Ludwigs, Chem. Rev. **110**(8), 4724–4771 (2010)
32. V.D. Pokhodenko, V.A. Krylov, Theor. Exp. Chem. **30**(3), 91–105 (1994)
33. C.A. Cutler, A.K. Burrell, D.L. Officer, C.O. Too, G.G. Wallace, Synth. Met. **128**(1), 35–42 (2002)
34. S. Alkan, C. Cutler, J. Reynolds, Adv. Funct. Mater. **13**(4), 331–336 (2003)
35. B.A. Miller-Chou, J.L. Koenig, Prog. Polym. Sci. **28**(8), 1223–1270 (2003)
36. R.A. Hayes, J. Appl. Polym. Sci. **5**(15), 318–321 (1961)
37. C.M. Hansen, *Hansen Solubility Parameters: a User's Handbook* (CRC Press, 2000)
38. T. Lindvig, M.L. Michelsen, G.M. Kontogeorgis, Fluid Phase Equilib. **203**(1–2), 247–260 (2002)
39. F. Machui, S. Langner, X. Zhu, S. Abbott, C.J. Brabec, Sol. Energy Mater. Sol. Cells **100**, 138–146 (2012)
40. P.S. Murray, S.F. Ralph, C.O. Too, G.G. Wallace, Electrochim.Acta **51**(12), 2471–2476 (2006)
41. X. Du, Z. Wang, Electrochim. Acta **48**(12), 1713–1717 (2003)
42. D. Maddison, J. Unsworth, Synth. Met. **30**(1), 47–55 (1989)
43. E.A. Bazzaoui, S. Aeiyach, P. Lacaze, J. Electroanal. Chem. **364**(1–2), 63–69 (1994)
44. S. Mu, S.-M. Park, Synth. Met. **69**(1–3), 311–312 (1995)
45. Y. Yuan, S. Adeloju, G. Wallace, Eur. Polym. J. **35**(10), 1761–1772 (1999)
46. J.M. Charlesworth, A.C. Partridge, N. Garrard, J. Phys. Chem. **97**(20), 5418–5423 (1993)
47. P. Pichanusakorn, P. Bandaru, Mater. Sci. Eng. R **67**(2–4), 19–63 (2010)
48. L. Bao-Yang, L. Cong-Cong, L. Shan, X. Jing-Kun, J. Feng-Xing, L. Yu-Zhen, Z. Zhuo, Chinese Phys. Lett. **27**(5), 057201 (2010)
49. B. Zhang, J. Sun, H.E. Katz, F. Fang, R.L. Opila, ACS Appl. Mater. Interfaces **2**(11), 3170–3178 (2010)
50. D.K. Taggart, Y. Yang, S.-C. Kung, T.M. McIntire, R.M. Penner, Nano Lett. **11**(1), 125–131 (2011)

Chapter 8
Atomic Layer Deposition of Metal Oxides

With a current record conversion efficiency of 12.3 %, dye-sensitized solar cells (DSSCs) cannot quite yet compete with their conventional solid-state counterparts. Nevertheless, because they are based on low-cost and abundant titanium dioxide, they are considered promising candidates for future solar energy harvesting applications [1]. Double-gyroid structured titania films of appropriate thicknesses are expected to significantly enhance the efficiency of DSSCs due to the nanomorphology's advantageous properties [2–4]. These include the high accessible surface area, which is essential for a high dye loading and effective charge separation. The monolithic nature of the metal oxide network is important for an efficient charge transport within the nanostructure.

Unfortunately, the previously presented templated-assisted nanostructuring strategies based on electrochemical deposition have only a limited applicability for DSSCs. Electroplating of metallic titanium is considered infeasible, preventing the post-deposition thermal oxidation approach to nanostructured titania, see Chap. 6 [5]. The fabrication of double-gyroid-structured titania via the templated electrodeposition of hydrated TiO_2 has been demonstrated. However, deposits with thicknesses exceeding 1 μm suffer from stress-induced delamination during thermal dehydration, which limits the efficiency gain when applied in DSSCs [3, 4]. Therefore, the objective of this last part of the study was to investigate an alternative titania deposition method for replicating the gyroidal templates which does not involve a hydrated intermediate.

In collaboration with **Pedro Cunha** we focused our joint efforts on the refilling of the mesoporous organic scaffolds via gas phase atomic layer deposition (ALD) of metal oxides, especially titania. ALD is based on the self-limiting and alternating reaction of the surface to be coated with at least two precursors in a sequential manner, see Fig. 8.1. Importantly, this allows the manufacture of uniform surfaces coatings of high purity with sub-nanometer thickness precision. Although conceptually straightforward, coating and filling the voided nano-channels of double-gyroid-structured organic films proved to be extremely challenging. The main reason for this is the inherently small pore diameter which decreases even further with progressing

M. R. J. Scherer, *Double-Gyroid-Structured Functional Materials*,
Springer Theses, DOI: 10.1007/978-3-319-00354-2_8,

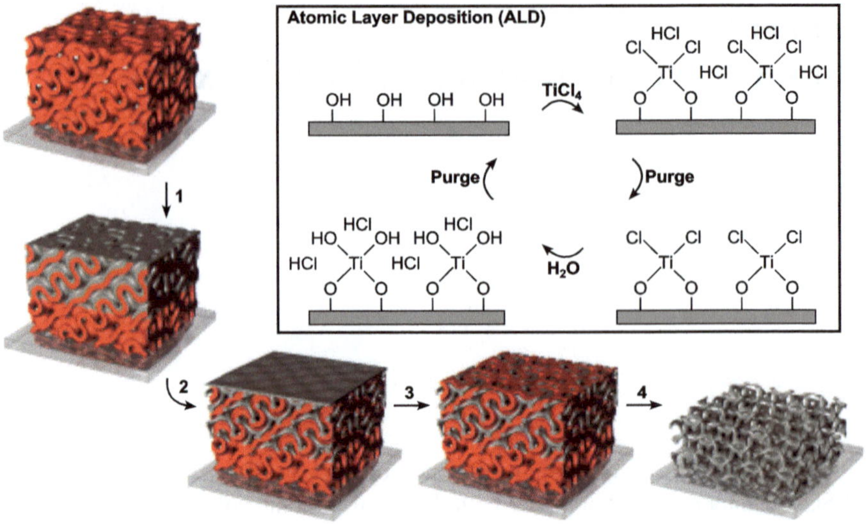

Fig. 8.1 Refilling of the voided nano-channels of an organic DG template via atomic layer deposition of titania, which is illustrated in the inset. *1* Chemical modification of the styrenic polymer scaffold to introduce functional surface groups and improve the thermal stability. This enables the uniform nucleation of the ALD growth *2* Ideally, the nano-channels are gradually filled until a non-porous layer is formed at the free-surface. *3* This layer is removed by reactive ion etching. *4* Finally, the combined organic/inorganic deposit is calcinated to remove the template and ideally, crystallize the titania

deposition. Thereby, the uniform exposure of the nano-channel walls to precursor, as well as the extraction of excess precursor and by-product becomes more difficult. This was complicated by the fact that the as-prepared styrenic templates do not exhibit adequate surface functional groups essential for a homogenous nucleation of the ALD growth, nor the temperature stability required for an ideal ALD process. Here, a newly established protocol is presented, that allows a simultaneous surface chemistry modification and a thermal stability improvement of the styrenic scaffold without altering its nanostructure.

Even though a stability increase by more than 30 °C was achieved, the tested maximal deposition temperature of 100 °C is considered non-ideal for the refilling of mesopores with TiO_2 using $TiCl_4$ and H_2O. The reduced reactivity, volatility, and mobility of the precursors and by-products at this relatively low temperature results in nonuniform growth and an incorporation of by-products and/or unreacted ligands [6]. We believe that the limited filling fraction, the observed chlorine contamination, and therefore the missing crystallinity are responsible for the disappointing performance of the prepared TiO_2 nanostructures in DSSCs.

Despite committing a respectable amount of time to this project, we have not yet succeeded to obtain crystalline double-gyroid-structured TiO_2 films. Based on the here represented results we are planing to test a homebuilt ALD which is currently under construction and was especially designed for high precursor concentrations and long exposure times. Additionally, we will try to improve the thermal stability of the templates even further. However, I am no longer fully convinced that replicating micrometer thick double-gyroid templates with an initial pore diameter of only 11 nm via ALD is experimentally feasible.

8.1 Introduction to Atomic Layer Deposition and the Chemical Modification of Poly(Styrene)

8.1.1 Atomic Layer Deposition

ALD is a thin film coating technique based on the self-terminating gas-solid reaction of gaseous precursors with the surface to be coated, characterized by the distinct feature that film growth takes place in a cyclic manner. [7]. A typical (sub)monolayer growth cycle consists of the strictly sequential delivery and self-saturated chemisorption of at least two precursors from the gas phase [8]. Importantly, between sequential precursor pulses the reaction chamber is evacuated or purged with an inert gas to remove non-reacted precursor molecules and the by-product of the chemisorption reaction. Figure 8.1 schematically illustrates the ALD process of TiO_2 using $TiCl_4$ and H_2O as precursors which proceeds via the formation of hydrochloric acid as by-product. The reaction of $TiCl_4$ is self-limited in the sense that once all OH groups have been consumed or are not accessible due to steric hindrance, the excess precursor will remain non-reacted with no further chemisorption occurring during extended exposure times. In an ideal ALD scenario a subsequent purge carries all by-product and the unused precursor molecules away. During the second half-cycle, the surface is exposed to H_2O molecules, which are chemisorbed by reacting with the remaining chlorine ligands, thereby forming the next layer of OH groups. Purging concludes a complete ALD cycle. A direct consequence of this saturated, self-limited surface reaction is the constant growth rate with atomic level control and the conformality of three-dimensional coatings which no other thin film technique can achieve.

A large variety of materials can be deposited via ALD, including pure metals, organic polymers, hybrid materials, and metal ceramics, such as oxides, nitrides, sulfides, selenides, and tellurides [7, 9]. Depending on experimental settings, precursors have to fulfill special requirements regarding their volatility, reactivity, and stability. These can be altered by exchanging the precursor ligands. For example, common titania precursors are titanium tetrachloride, titanium isopropoxide or tetrakis(dimethylamino)titanium [10, 11]. Accordingly, several oxygen source are available to prepare metal oxides, including water, hydrogen peroxide, oxygen, ozone, and oxygen radicals.

The deposition process is governed by the precursor volatility, stability, and reactivity which mainly depend on the substrate/reactor temperature for a given reactor pressure, as illustrated in Fig. 8.2. The ideal process window, the so-called ALD window, in which growth is saturated and insensitive to overexposure with precursor.

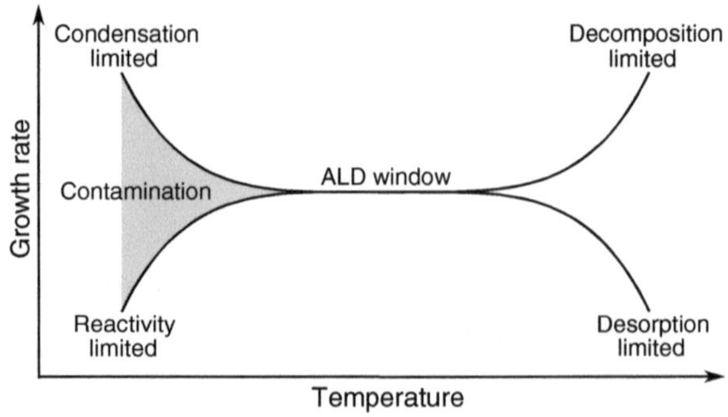

Fig. 8.2 Dependence of the ALD growth rate on the deposition temperature. The ALD window is characterized by a stable deposition rate reflecting an ideal precursor volatility, reactivity, and stability in this temperature range

At temperatures above the ALD window, the precursor molecules typically undergo thermal decomposition resulting in an uncontrolled deposition. Although ALD is mainly applied to inorganic substrates which can resist the often elevated temperatures of the ideal ALD window, there is a growing interest in low temperature ALD on temperature sensitive organic substrates or templates [12–14]. However, ALD performed at temperatures compatible with organic substrates is often subject to a low deposition rate and an incorporation of non-reacted ligands due to a reduced reactivity and/or suffers from an uncontrolled multilayer deposition involving physisorbed precursor caused by a decreased volatility of the latter. Furthermore, a decreased volatility of the by-product may cause unwanted contamination of the ALD deposit.

Furthermore, the deposition temperatures significantly influences the growth mode of the deposit. For example, at temperatures above 165 °C crystalline TiO_2 with a pronounced surface roughness is deposited, whereas smooth amorphous coatings, which are ideal when attempting the refilling of mesopores, are obtained below this temperature [6, 15]. However, crystallization was found to be suppressed in very thin coatings or by interactions with the substrate surface. The deposition behavior at temperatures between 80 and 165 °C was studied in more detail by Aarik et al. and Triani et al. [6, 14, 15]. In this temperature range they find an increasing growth rate and mass gain with decreasing temperature which is most probably due to the retention of physisorbed precursor molecules on the growth surface and/or incorporation of ligands and by-product. This in turn leads to uncontrolled deposition and may cause pore clogging at the free-surface when refilling mesoporous templates. Indeed, the detectable amount of chlorine incorporated in titania coatings increases significantly with a decreasing deposition temperature. Based on these facts, a deposition temperature of around 150 °C should be considered optimal for replicating mesoporous scaffolds with TiO_2 using $TiCl_4$.

Area-selective ALD is possible, since deposition is based on chemisorption which only occurs in regions of the surface where reactive sites, such as OH and NH_2, are present [8]. Typically, monolayers of silane with chemically inert tails are used to passivate native functional surface groups. Similarly, standard (fluorinated) poly(styrene) does not have any functional groups, resulting in a slow initiation of the deposition and island growth on such polymer substrates. Indeed, Puttaswamy et al. found a delayed deposition of Al_2O_3 on untreated poly(styrene) using trimethylaluminium (TMA) and water as precursors [8, 16]. Only after 40 ALD cycles they could detect aluminium with X-ray photoelectron spectroscopy (XPS). The high surface roughness of the alumina coatings prepared on smooth spin-coated polymer films suggests island growth caused by a low nucleation density. This nonuniform growth will most likely circumvent the monolithic replication of such mesoporous polymer templates with TiO_2, especially since $TiCl_4$ is considered less reactive than TMA which may decrease the nucleation density even further. On the other hand, Wilson et al. observed an immediate mass increase also during the first ALD cycles [17]. Furthermore, precursors such as TMA can diffuse into organic materials causing subsurface deposition with depths of several tens of nanometers [18]. Especially in nanostructured polymeric templates this can lead to a chemical modification of the polymeric matrix.

The fact that ALD is based on a self-terminating gas-solid reaction yielding excellent deposition conformality allows to coat high aspect ratio nanostructures, including colloidal arrays, anodized alumina and track etched poly(carbonate) membranes [19, 20].

The reader is referred to the comprehensive reviews from Knez et al. and Detavernier et al. on this topic [21, 22]. Sufficiently long pulse and purge steps are required to achieve saturated chemisorption and complete removal of unused precursor or by-product, respectively. Since the dimensions of nanostructured templates are comparable or smaller than the mean free path of the gaseous precursor and by-product molecules the mass transport is considered to be governed by Knudsen diffusion [23]. Indeed, good agreement between Monte Carlo simulations based on a Knudsen diffusion model and the corresponding experimental results was found [24]. Karuturi et al. studied TiO_2 ALD in colloidal templates at 70 °C [25, 26]. Although the coating of the high aspect ratio colloidal templates was Knudsen diffusion limited they reported non-ideal ALD behavior, such as nonuniform coatings that varied with precursor concentration and exposure time. Gordon et al. proposed the following analytical expression for the minimal required exposure time t to conformally coat a cylindrical nanopore with a certain aspect ratio a, which assumes an ideal ALD process with a 100 % sticking probability of precursor molecules on reactive surface sites and zero residence time for multilayer adsorption [27]

$$t = \frac{S}{P}\sqrt{2\pi mkT}\left(1 + \frac{19}{4}a + \frac{3}{2}a^2\right), \tag{8.1}$$

where P is the precursor partial pressure, S is the saturated surface density, m is the mass of the precursor molecules, k is the Boltzmann constant and T is the temperature. However, this model was developed and tested for cases where the pore diameter is large compared to the diameter of the precursor molecules, hence taking only the aspect ratio into account seems sufficient. The results obtained by Kucheyev et al.

suggest, that the coating depth of silica aerogel monoliths with an initial pore diameters of 18 or 27 nm is limited not only by Knudsen diffusion of precursor molecules into the pores, as believed previously, but also by other processes such as the interaction of precursor and reaction product molecules with pore walls [28]. They are of the opinion, that even a relatively small residence time of physisorption would dramatically reduce gas diffusion through mesopores. However, the penetration depth was found not to increase with exposure time as it would be expected for this case. They argued further, that confinement effects could aid precursor decomposition even at temperature where the liquid phase precursor typically is stable.

8.1.2 Post-synthesis Chemical Modification of Poly(styrene)

In the literature several methods to chemically modify poly(styrene) have been reported. The interest in the surface functionalization stems from the need to improve cell adhesion on PS petri-dishes for medical and biological applications, while PS cross-linking techniques were investigated for potential application as lithography resist. However, only two of the numerous reported techniques proved to be effective and suitable for the mesoporous styrenic templates, namely exposure to ozone or plasma ignited in several atmospheres, such as nitrogen, air, oxygen, and water vapor.

Exposure to ozone or plasma is known to render poly(styrene) surfaces hydrophilic in a controlled manner. Both techniques partially degrade the polymer, producing low molecular weight fragments which are easily washed away to leave a stable modified surface. Generally X-ray photoelectron spectroscopy is used to analyze the atomic composition of the modified surface which comprises species containing C–O (peroxide, alcohol, and ether), C=O (aldehyde, ketone), and O=C–O (ester, carboxylic acid) groups, of which only the hydroxyl containing species react with $TiCl_4$ [29]. Unfortunately, the XPS data presented in the literature does not provide any information concerning the fraction of hydroxyl containing species. While the oxidized groups of plasma-treated films are know to contain a high ratio of the latter, the same is believed not to be true for ozone-treated films [30–32].

Due to the high reactivity and short lifetime of radicals, plasma induced surface modification is only limited applicable to high aspect ratio nanostructures. In contrast, ozone molecules are more stable and far less reactive, and as such ideal for the gas phase surface modification of mesoporous poly(styrene) by Knudsen diffusion [33, 34]. According to the Criegee mechanism, the reaction of alkenes and ozone to secondary ozonides (1,2,4-trioxolanes) can be described as a cycloreversions of primary ozonides (1,3,3-trioxolanes) involving aldehydes or ketones as decisive intermediates. Although the secondary ozonide is more stable than the intermediates it also has a finite lifetime. During subsequent work-up strong oxidizing or reducing agents are used to transform the secondary ozonide into carboxylic acids/ketones, aldehydes/ketones, or alcohols, see Fig. 8.3.

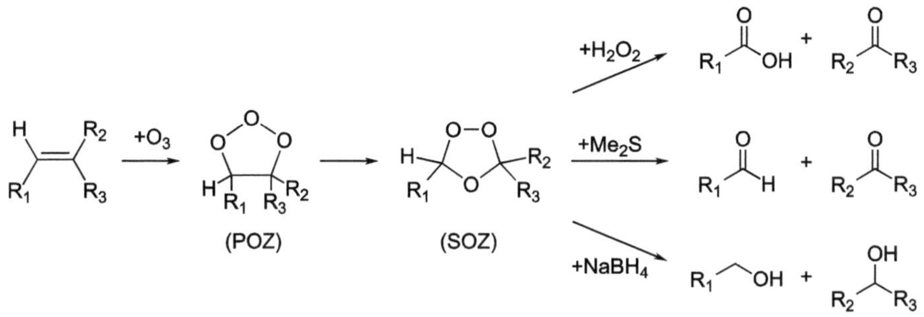

Fig. 8.3 Ozonolysis allows the cleavage of alkene double bonds. According to the Criegee mechanism the primary ozonide (*POZ*) is rapidly transformed into the more stable secondary ozonide (*SOZ*). Depending on the work-up, different products may be isolated. Oxidative work-up with hydrogen peroxide leads to carboxylic acids/ketones, while reductive work-up with either dimethyl sulfide or sodium borohydride gives aldehydes/ketones or alcohols, respectively

Table 8.1 Chemicals and materials used in this part of the study

Chemical	Purity [%][a]	Abbreviation
2-Amino-2-hydroxymethyl-propane-1,3-diol	99 (SA)	TRIS
α-Picoline borane	95 (SA)	–
CELLSTAR petri dishes (lids only)	–(GBO)	–
Diethylzinc	– (SA)	DEZ
Dimethyl sulfide	99 (SA)	Me_2S
Ethanolamine	99 (SA)	–
Hydrazine hydrate	50–60 (SA)	–
Hydrogen peroxide	30/50 (SA)	–
Methanol, laboratory reagent grade	– (FS)	–
Nitric acid, laboratory reagent grade	70 (FS)	–
Oxygen	99.999 (BOC)	–
Sodium borohydride	98 (SA)	–
Sulfuric acid, laboratory reagent grade	95 (FS)	–
Titanium isopropoxide	95 (SA)	TIP
Titanium tetrachloride	99 (SA)	–
Toluene, anhydrous	99.8 (SA)	–
Trifluoroethanol	99 (SA)	–
Trimethylaluminium	97 (SA)	TMA
Water, deionized 18 MΩ	–	DI water

[a] Suppliers are given in brackets: Fisher Scientific, FS; Sigma Aldrich, SA; Greiner Bio-One, GBO

8.2 Experimental Methods

8.2.1 Materials

The chemicals and materials listed in Table 8.1 were used as received unless stated otherwise.

8.2.2 Atomic Layer Deposition

The ALD system used was a Beneq TFS200. Numerous different recipes were tested to replicate the double-gyroid templates. Here, the discussion is limited to the most successful pulse-hold recipe consisting of the following four steps with an increasing hold duration.

	Cycles	Precursors	Pulse/hold/purge [s]
Step 1 (optional)	1	Precursor 1 and 2	15/120/240
Step 2	5	Precursor 3 and 4	10/0/60
Step 3	20	Precursor 3 and 4	10/120/240
Step 4	30	Precursor 3 and 4	10/240/480

Since $TiCl_4$ merely reacts with hydroxyl groups, more reactive precursors such as TMA or DEZ are used for 'activation' of other functional surface groups that do react with the latter in the optional Step 1. In the following, this ALD recipe for refilling the double-gyroid templates will be denoted as MetalOxide $\left(\substack{\text{Precursor 1}\\ \text{Precursor 2}}\middle|\substack{\text{Precursor 3}\\ \text{Precursor 4}}\middle|\ \text{Temperature}\right)$.

8.2.3 Ozonolysis

Poly(styrene) samples were exposed to a constant flow of ozone (ozone generator BMT 802N operated at 100 % power) at room temperature for 15 min. Thereafter, the samples were annealed for 24 h at 60 °C in air.

8.2.4 SEM Sample Preparation

Calcinated samples were fractured and the cross-sections were imaged without any further preparation steps. As-deposited (non-calcinated) ALD films were fractured and then etched with oxygen plasma for 2 min at 100 % power to free the cross-sections from the polymer template.

8.2.5 *Fourier Transform Infrared and Energy-Dispersive X-ray Spectroscopy*

The untreated lids of CELLSTAR petri dishes, which exhibit a high water contact angle (>90°), were used as model system for the various surface functionalization methods tested. The effect of the treatment on the samples was analyzed by means of attenuated total reflection Fourier transform infrared spectroscopy (ATR-FTIR) and water contact angle measurements. Furthermore, the ability of the modified surface to initiate an uniform ALD growth was quantified by energy-dispersive X-ray spectroscopy (EDX) using an acceleration voltage of 15 keV. Since the EDX signal for titanium is very weak and the low-energy peaks overlap with the oxygen peak, zinc oxide was chosen as ALD material instead. In some cases the petri dishes were exposed to a TMA pulse prior to the ZnO ALD which was performed at 65 °C.

	Cycles	Precursors	Pulse/hold/purge [s]
Step 1 (optional)	1	TMA/H_2O	0.5/0/2
Step 2	40	DEZ/H_2O	0.5/0/2

8.3 Results and Discussion

For the refilling of the double-gyroid templates via electrochemical deposition a fully porous film/substrate interface was crucial while only a few pores at the free-surface are needed for electrolyte infiltration. Contrary, ALD requires a highly porous free-surface for efficient gas diffusion into the mesopores, while the porosity of the film/substrate interface is not relevant. Throughout this last part of the study, P(F)S49-b-PLA192$\Big|^{174\,^\circ\mathrm{C}}_{0.33\mathrm{h}}$ templates exhibiting a double-gyroid-structured and therefore porous free-surface were used, see Sect. 4.3.5.

8.3.1 *ALD on Untreated Templates*

Zalusky et al. showed that mesoporous poly(styrene) scaffolds exhibit a limited thermal stability which results in the loss of the nanostructure features at temperatures well below the glass transition temperature of about 104 °C [35]. This is not surprising considering that the glass transition temperature is a concept that applies to bulk material. Similarly, voided DG-structured films prepared from P(F)S49-b-

$\text{PLA192}\Big|^{174\,^\circ\text{C}}_{0.33\text{h}}$ were affected by pore clogging and structure loss at temperatures exceeding 70 °C, hence ALD on untreated templates was carried out at 65 °C.

The chemical group connecting the two blocks of microphase separated P(X)S-*b*-PLA is expected to lie at the interface between the incompatible P(X)S and PLA domains. Zalusky et al. demonstrated that the connecting groups decorate the pore walls after selective PLA removal [35, 36]. They further estimated the corresponding surface density to be 0.25 nm^{-2}. Since HEBIB was used as bifunctional initiator in this study, a functional carboxylic acid end-group should be present at the nano-channel surface. Therefore, it was expected that the nucleation density inside the mesoporous styrenic templates is sufficiently high to yield conformal coatings.

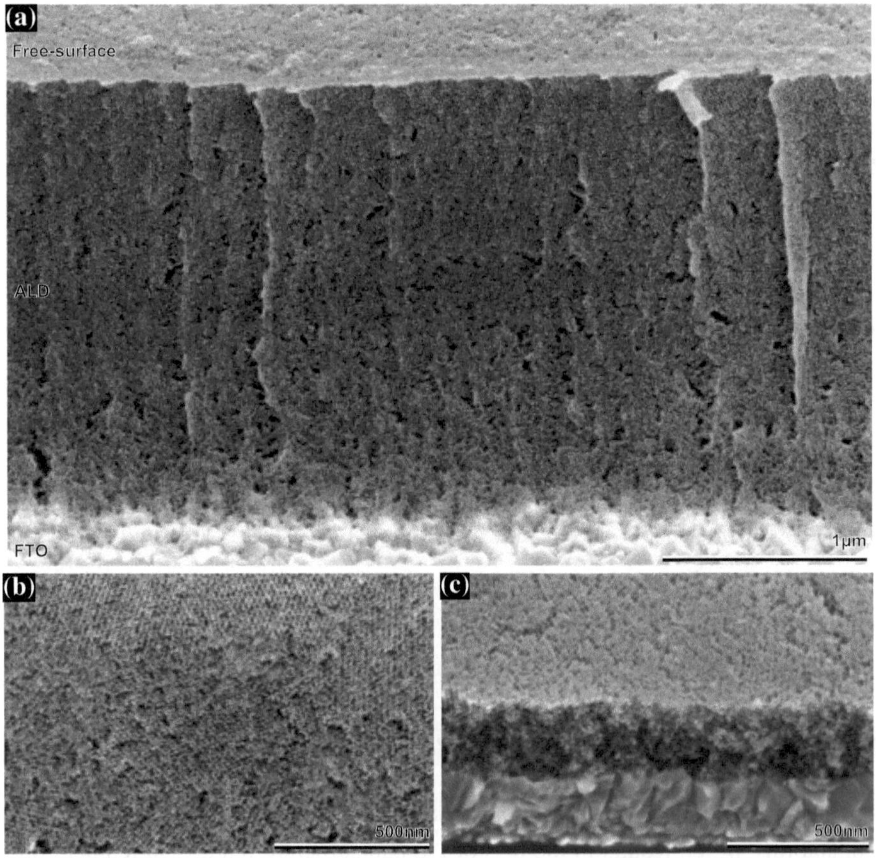

Fig. 8.4 ALD of TiO_2 $\left(\frac{-}{-}\Big|^{TiCl_4}_{H_2O}\Big|65^\circ C\right)$ on untreated templates. **a** Cross-sectional view of the titania replica after partial template removal with plasma. Note the grainy and perforated island growth. **b** Magnified view of (**a**) (**c**) Collapsed film after calcination at 500 °C

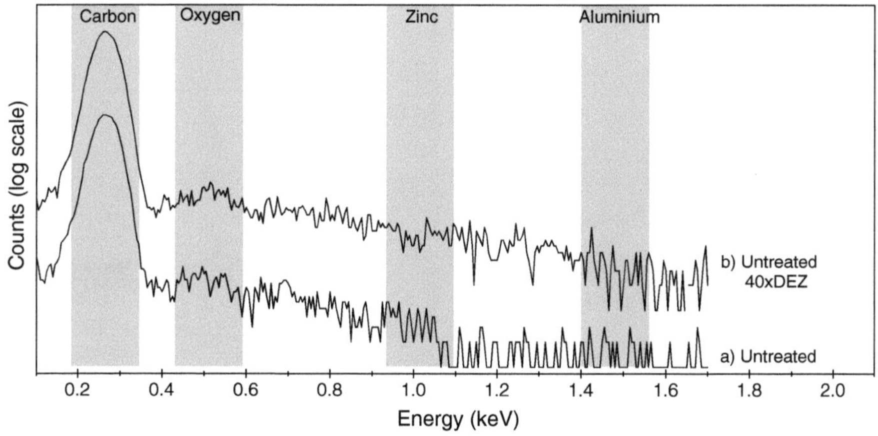

Fig. 8.5 EDX spectra of untreated poly(styrene) petri dishes without and with ALD coating (40 cycles DEZ/H_2O). For comparison, the carbon peak was normalized. There is no evident difference between the two samples

However, the nucleation density of both TiO_2 $\left(\overline{-}\middle|^{TiCl_4}_{H_2O}\middle|65\,°C\right)$ and Al_2O_3 $\left(\overline{-}\middle|^{TMA}_{H_2O}\middle|65\,°C\right)$, on untreated templates was unsatisfactorily low resulting in grainy and perforated island growth, as shown in Fig. 8.4a, b. On calcination, the initially several micrometer thick replica collapsed and was reduced to a thin film with no structural order (<300 nm), see Fig. 8.4c. Attempts to improve the quality of the replica by modifying the ALD recipe described in Sect. 8.2.2 were unsuccessful. Indeed, no zinc signal was observed in EDX spectra for poly(styrene) petri dishes exposed to 40 ALD cycles of DEZ/H_2O, see Fig. 8.5. Importantly, 40 cycles of DEZ/H_2O are sufficient to obtain a zinc signal on silicon or plasma-treated samples, see Fig. 8.6. We attribute the unsatisfactorily results to the absence of a sufficiently high surface density of functional groups rather than to the low deposition temperature. This hypothesis is supported by the fact that monolithic replica were prepared at this temperature on plasma-treated templates.

8.3.2 ALD on Plasma-Treated Templates

It is known that very short exposures to an oxygen plasma are sufficient to render polymer surfaces hydrophilic and to introduce functional surface groups, such as hydroxyl containing species. The highly reactive plasma radicals are not expected to penetrate the poly(styrene) bulk material but only react with the outermost surface layer. Indeed, the ALD nucleation and growth on the petri dishes was significantly improved by an air plasma treatment with a Diener MRC 100 for less than 0.5 s. While an oxygen signal was observed by EDX and the water angle was reduced to

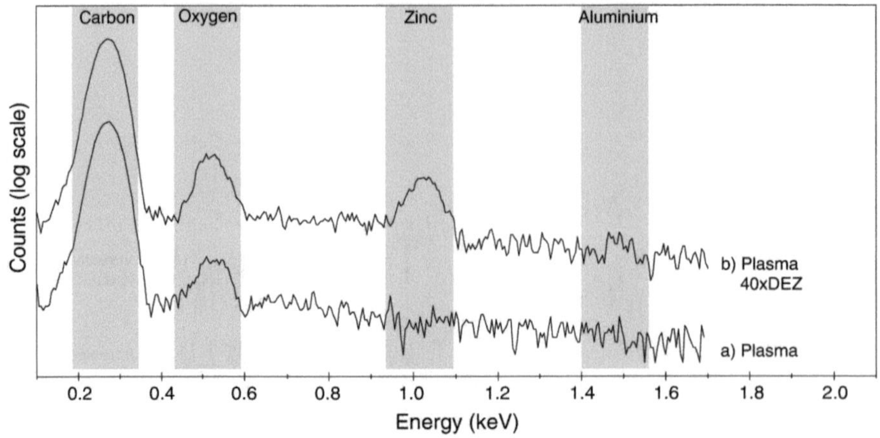

Fig. 8.6 EDX spectra of 0.5 s air plasma-treated petri dishes without and with ALD coating. In contrast to untreated poly(styrene) a well resolved Zn peak is observed

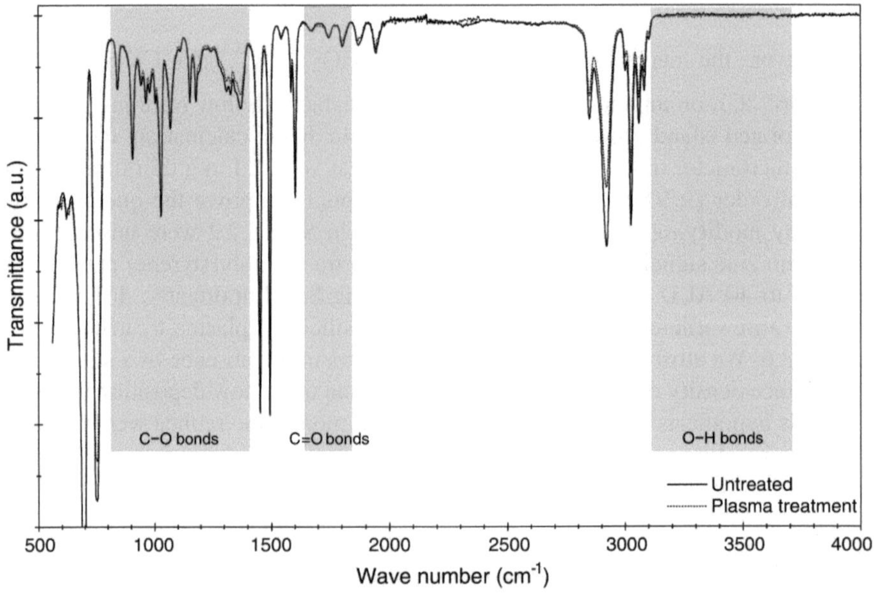

Fig. 8.7 ATR-FTIR spectra of an untreated and air plasma-treated (0.5 s) petri dish. For comparison, the peak at 700 cm^{-1}, which is associated with aromatic and vinyl C–H bonds, was normalized. Besides a general increase of the peak hight no additional peaks were observed

0 °, no hydroxyl groups were detected by ATR-FTIR, compare Figs. 8.6 and 8.7. This is not surprising considering that EDX probes the atomic composition of the surface, while the photons used for FTIR penetrate the bulk material.

Next, double-gyroid templates were plasma-treated. In order to minimize polymer etching and excessive heating that may cause melting of the nanostructure during prolonged treatment ranging from 10 min to 10 h, a low generator power (40 %) and low gas pressure (1 mbar) were chosen. Additionally the samples were placed underneath a microscope slide to simulate a remote plasma. SEM images of TiO_2 $\left(\frac{-}{-}\Big|\frac{TiCl_4}{H_2O}\Big|65\,°C\right)$ deposits show compact and monolithic replica of the double-gyroid template, see Fig. 8.7. The distinctly sharp transition from the replicated to the underlying non-refilled part of the template is attributed to the limited penetration depth of the plasma. Since the plasma radicals are highly reactive their lifetime limits the diffusion into the mesoporous film. The difference in surface density of ALD nucleation site decorating the nano-channel that lay within and outside the plasma diffusion depth caused the distinctly different ALD initiation and growth speed across the template. A more continuous transition is expected when the ALD precursor diffusion is the limiting factor, as will be discussed in Sect. 8.3.4.

The maximal refilling depth never exceeded 1 μm, regardless of the plasma treatment time, see Fig. 8.8. Moreover, plasmas ignited in several other atmospheres, such as oxygen, nitrogen, and water vapor, did not improve the penetration depth. This also shows, that plasma etching of a nonporous layer covering the free-surface of a template, which is meant to be surface functionalized by another method than plasma, is possibly not feasible, since the plasma-etched free-surface most likely will exhibit a higher surface density of functional groups which in turn will lead to pore clogging.

As discussed in Chap. 2, the complete refilling of the gyroidal nano-channels via ALD is not possible because of the pinch-off problem. Additionally, the finite molecular radius of the precursor molecules, e. g. $r_{TiCl_4} = 0.32\,\text{nm}$ and $r_{H_2O} = 0.16\,\text{nm}$, will consequently cause the formation of nanotubular replica, see Fig. 8.8b [37]. However, the unchanged low thermal stability and the therefore low deposition temperature caused high levels of chlorine contamination (≈15 %) which were only reduced marginally during calcination, as measured by EDX.

8.3.3 *Unsuccessful Attempts to Chemically Modify Poly(styrene)*

After identifying missing surface functional groups as the main reason for the grainy deposit and the low deposition temperature as cause for the high level of chlorine doping, a number for other methods claimed to chemically modify poly(styrene) were tested. These included nitration with dissolved nitrites or gaseous nitrogen dioxide [38–42], sulfonation with concentrated sulfuric acid [43], oxidation with sodium hydroxide or potassium permanganate [44], and photo-oxidation [30, 45–48]. Besides sulfonation and photo-oxidation, all remaining techniques failed to render the petri dishes hydrophilic or promote initiation of ALD growth. Unfortunately, both successful methods were found to destroy the nanostructured poly(styrene) templates (Fig. 8.9).

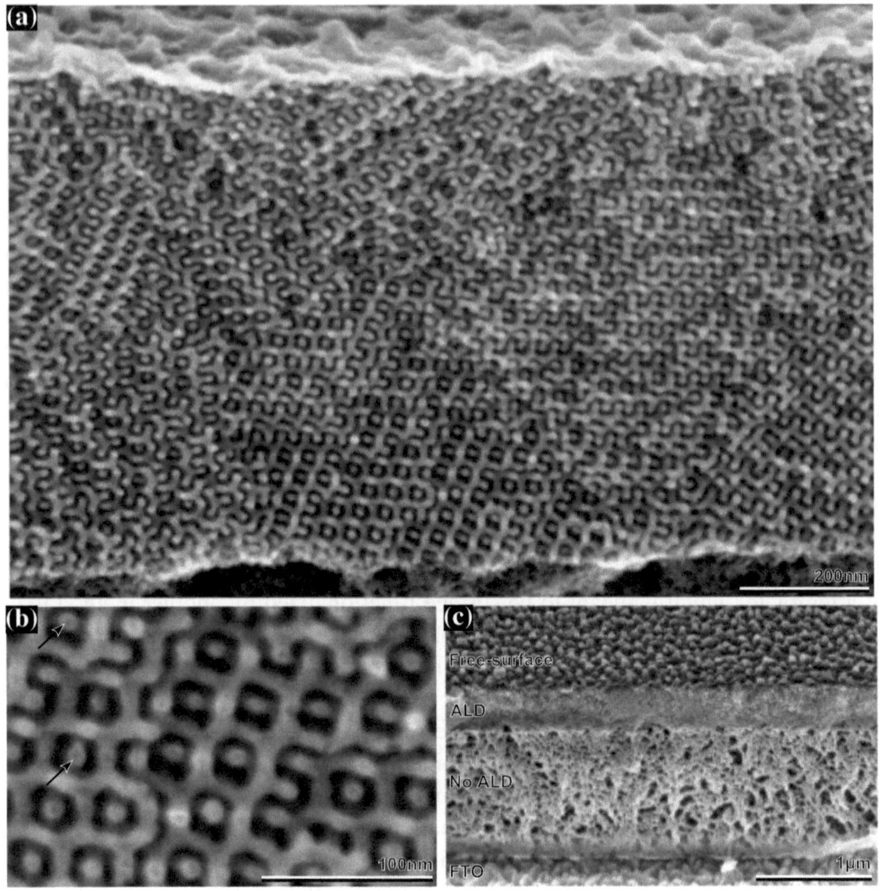

Fig. 8.8 ALD of $TiO_2\left(\frac{-}{-}\Big|\frac{TiCl_4}{H_2O}\Big|65\,°C\right)$ on air plasma-treated templates. **a** Magnified view of the refilled part of the cross-section shown in (**c**). The sharp transition from a compact and monolithic double-gyroid replica to the underlying unfilled film part was characteristic for plasma-treated templates. **b** Magnified view of (**a**). It is difficult to see, but the titania struts are hollow as expected. **c** Cross-sectional view of the titania replica after partial template removal with a plasma

8.3.4 ALD on Ozone-Treated Templates

The chemical modification of poly(styrene) by ozone has been reported in the literature [33, 34]. Indeed, a 15 min ozone exposure significantly reduced the water contact angle of the petri dishes and pronounced peaks associated with oxygen containing species, including hydroxyl groups, were detected, see Fig. 8.10a. Curiously, freshly ozone-treated surfaces did not promote an improved ALD nucleation neither for $TiO_2\left(\frac{-}{-}\Big|\frac{TiCl_4}{H_2O}\Big|65\,°C\right)$ nor for $ZnO\left(\frac{-}{-}\Big|\frac{DEZ}{H_2O}\Big|65\,°C\right)$. Further, the double-gyroid replica looked similar to the ones prepared using untreated templates, see Fig. 8.4.

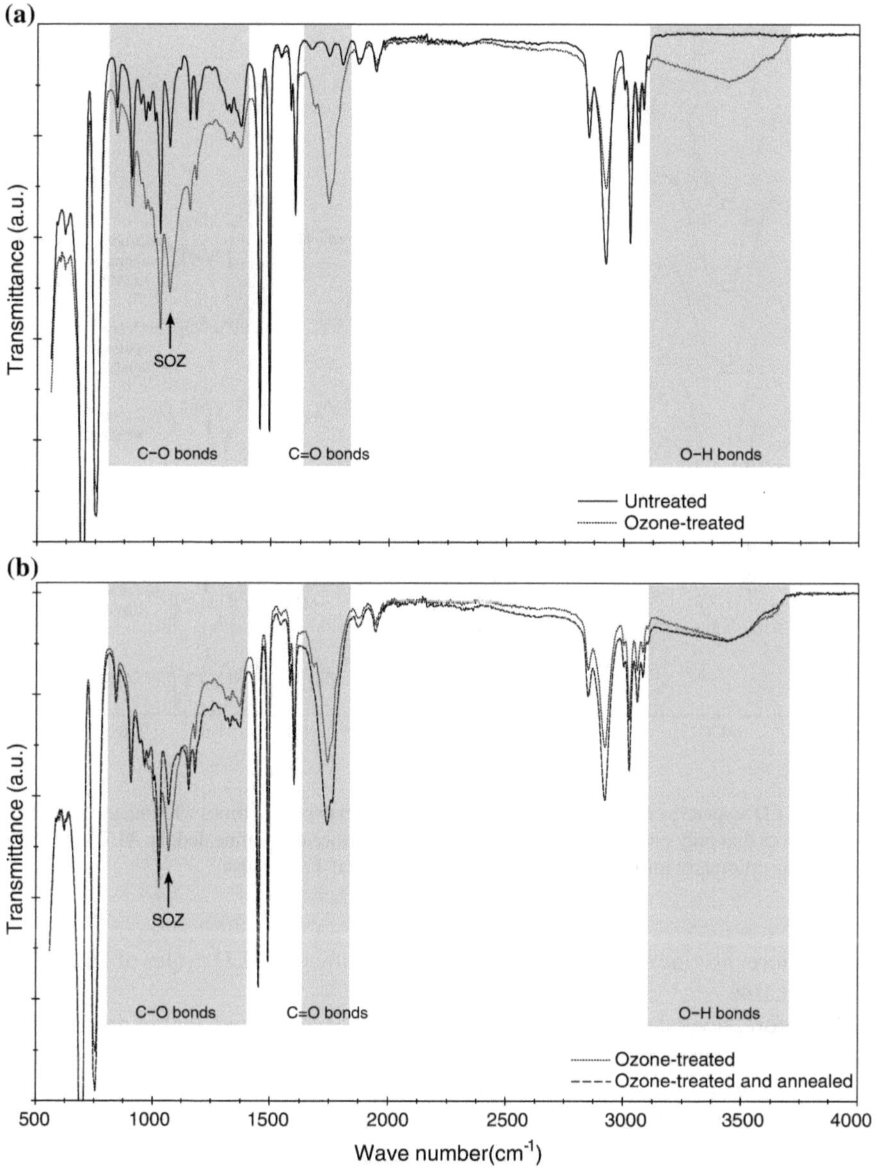

Fig. 8.9 ATR-FTIR spectra of ozone-treated petri dishes (15 min exposure time) [52]. **a** Juxtaposition of spectra measured for an untreated and a freshly ozone-treated sample. **b** Temperature annealing at 65 °C for 24 h in air resulted in an attenuation of measured signal arising from the secondary ozonide (SOZ), which was accompanied with an increase of the other oxygen containing species

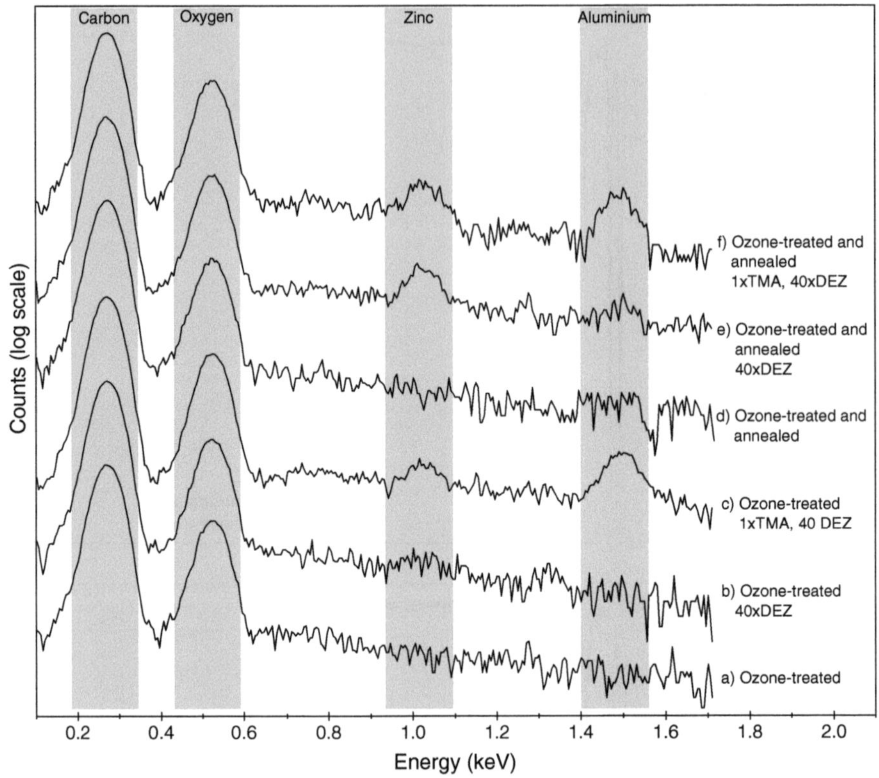

Fig. 8.10 EDX spectra of ozone-treated petri dishes (15 min exposure time) without and with ALD coating. Note that only once the ozone-treated sample were thermally annealed the ALD deposition of ZnO was efficiently initiated without the need of an initial TMA pulse

Furthermore, no zinc signal was detected by EDX after 40 ALD cycles of DEZ/H_2O, see Fig. 8.10b.

Therefore, several ozonolysis work-ups (Fig. 8.3) were tested, including the reductive work-up with sodium borohydride in trifluoroethanol at a pH of 13.7 as well as with Me_2S in methanol [49], and the oxidative work-up with H_2O_2. Especially, the reductive amination of aldehydes and ketones with TRIS, ethanolamine, and hydrazine using α-picoline-borane as reducing agent produced functional groups on petri dishes which are ideal for ALD using $TiCl_4$ or DEZ as precursors [50]. Unfortunately, the freshly ozone-treated nanostructured poly(styrene) dissolved during the wet chemistry based work-up. This finding is consistent with reports claiming, that low molecular products formed during ozonolysis are easily removed by soaking in water [29]. Luckily, TMA was found to react with the ozonides, aldehydes, and ketones formed during ozonolysis. One initial cycle of TMA/H_2O was found to be sufficient to decorate the ozone-treated styrenic surface with a layer of Al-OH containing species which constitutes an ideal surface for the subsequent ALD of either TiO_2 or ZnO, see Fig. 8.10c [51].

When analyzing the ATR-FTIR spectra of freshly ozone-treated petri dishes it became evident that the samples contained a high fraction of metastable secondary ozonides, which are known to thermally decompose. Therefore, it was attempted to increase the density of functional surface groups by thermally annealing the samples at 65 °C for 24 h in air. Indeed, the SOZ peak was significantly reduced during annealing, simultaneously the signal strength arising from other oxygen containing bonds was increased, see Fig. 8.9b. Although polymer samples modified in this way directly supported the ALD of ZnO removing the need for a TMA/H_2O cycle, this was not true for the ALD of TiO_2 using $TiCl_4/H_2O$ which still required an initial TMA or DEZ pulse.

Furthermore, the thermal stability of double-gyroid-structured P(F)S49-*b*-PLA192$|^{174\,°C}_{0.33\,h}$ templates was improved to over 100 °C by thermal annealing subsequent to ozonolysis, most probably due to cross-linking reactions. Note that unlike unsubstituted poly(styrene), poly(4-fluorostyrene) cannot be cross-linked by ultraviolet irradiation due to a stabilizing effect arising from the electron-withdrawing fluorine atoms. Although not tested, I believe that the chemical modification and cross-linking by ozonolysis and thermal annealing is highly promoted by the fact that P(F)S49-*b*-PLA192 only contains 30 % 4-fluorostyrene.

Figure 8.11 shows micrometer thick $TiO_2\left(\frac{TMA}{H_2O}\middle|\frac{TiCl_4}{H_2O}\middle|100\,°C\right)$ double-gyroid replica of freshly ozone-treated templates, which looked identical to $TiO_2\left(\frac{ZnO}{H_2O}\middle|\frac{TiCl_4}{H_2O}\middle|100\,°C\right)$ using annealed, ozone-treated scaffolds. The brighter appearance of the ALD replica closest to the free-surface in the SEM images indicates a compact and electrically conducting metal oxide deposit. On calcination the underlying darker regions vanished and the film thickness was typically reduced to below 1 μm, see Fig. 8.12d. They vanished in a sense that no compact or crumbled region was observed between the FTO and the intact part of the calcinated replica, suggesting that the ALD coating in these regions had to be extremely thin.

These results are somewhat puzzling, since it seems that the tested precursor exposure times should be sufficient to allow for a saturated chemisorption throughout the entire thickness of the templates. For example, consider a 5 μm thick double-gyroid template with an initial (i) pore diameter of about $d_i = 11$ nm that is gradually reduced to a final (f) diameter $d_f = 2$ nm. This translates to the following aspect ratios of $a_i = 455$ and $a_f = 2{,}500$, respectively. Together with the following assumptions, $S = \pi^{-1} r_{TiCl4}^{-2} = 3.1\,10^{18}\,m^{-2}$, $T = 373$ K, and $P = 1$ mbar (the partial pressure of $TiCl_4$ was probably much lower than this estimated value), Eq. 8.1 predicts a required minimal exposure time of 0.01 ms for a flat substrate ($a = 0$), $t_i = 3$ s for the initial cycle for a 5 μm thick double-gyroid template and $t_f = 94$ s for the final cycle, respectively. Even though the experimentally tested exposure times were significantly longer than these theoretically estimated values, only the first 1 μm of the nano-channels were refilled to a degree that they exhibited enough structural stability to endure the template removal during calcination.

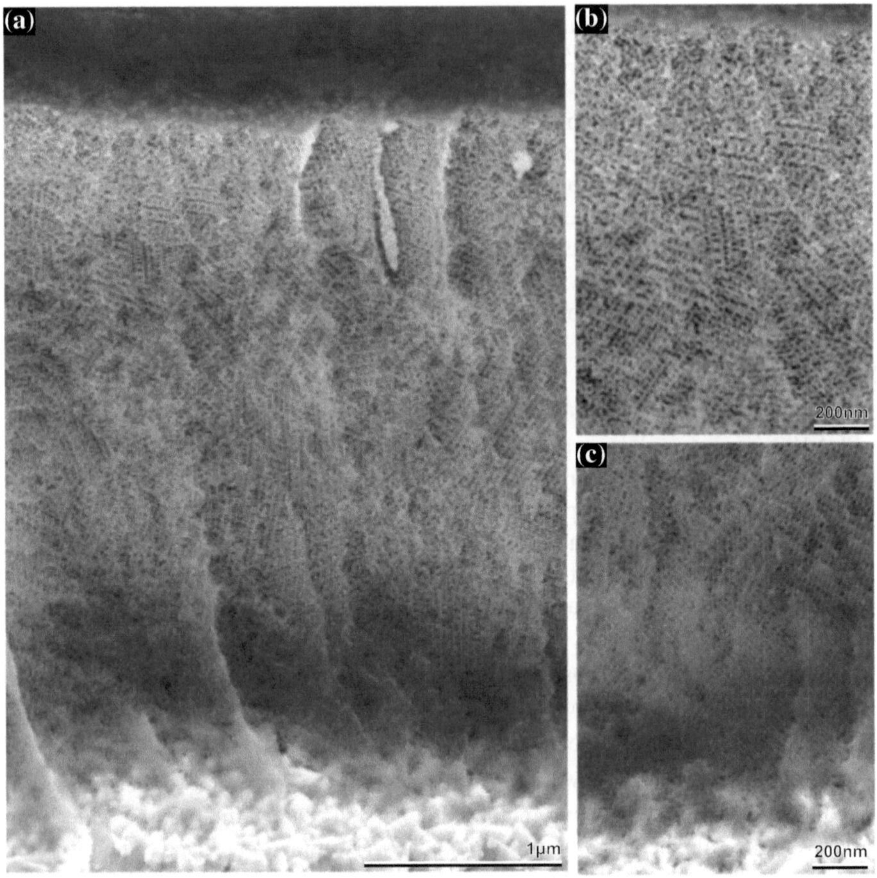

Fig. 8.11 ALD of $TiO_2\left(\frac{TMA}{H_2O}\middle|\frac{TiCl_4}{H_2O}\middle|100\,^{\circ}C\right)$ on ozone-treated templates. **a** Cross-sectional view of the titania replica after partial template removal with plasma. **b**,**c** Magnified view of the *bright top* and *darker bottom* part of replica shown in (**a**)

In contrast to the nanopores considered by Gordon et al., the initial pore diameter of the here tested double-gyroid templates was only 17 times the diameter of $TiCl_4$ [27]. This figure rapidly decreases as coating of the internal surface proceeds, thereby entering the adsorption thermodynamics regime where confinement effects start to become important. The adsorption energy in this case is expected to be one order of magnitude larger than for a nanostructured surface [53, 54]. Therefore surface diffusion, where the solute is adsorbed on the surface of the pore and hops from one site to another, should be considered rather than Knudsen diffusion [55]. This might significantly lengthen the time required to achieve saturated chemisorption as well as removal of the by-product and unreacted precursor. Furthermore, the ALD used was designed for coating flat 8-inch wafers with relatively short pulses and not for refilling high aspect ratio nanostructures. Especially the fact, that the reaction

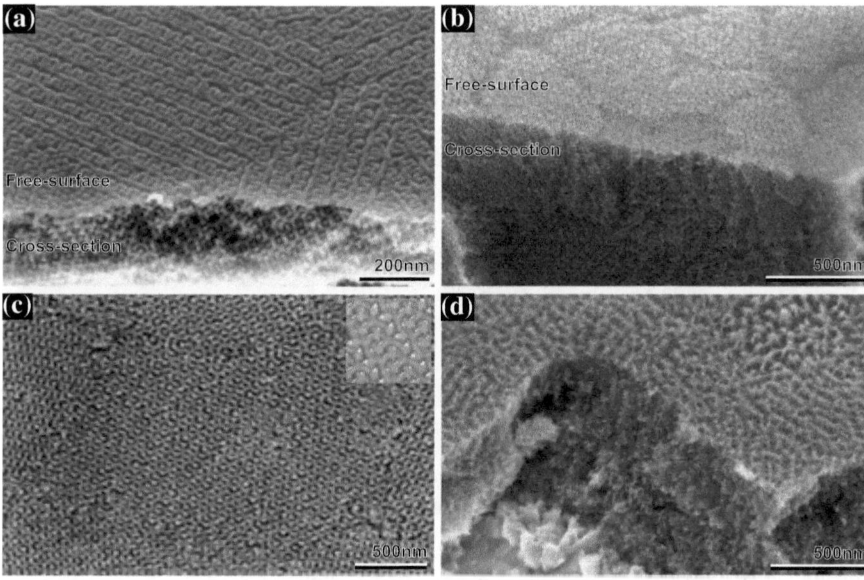

Fig. 8.12 ALD of $TiO_2\left(\frac{TMA}{H_2O}\middle|\frac{TiCl_4}{H_2O}\middle|100\,^\circ C\right)$ on ozone-treated templates. **a** The conformality of ALD results in a nonporous metal oxide layer covering free-surface. **b** Ideally, this layer should be removed by reactive-ion etching (RIE) before calcination. **c** Free-surface of (**a**) after RIE showing the (211) plane. **d** Film after calcination at 500 °C

chamber is not hermetically sealed off from the surrounding large vacuum chamber most likely caused a rapid decrease in precursor concentration during the hold step. This in turn would have undermined the purpose of the hold step and reduced the precursor exposure time to the actual pulse duration.

After washing the $TiO_2\left(\frac{DEZ}{H_2O}\middle|\frac{TiCl_4}{H_2O}\middle|100\,^\circ C\right)$ in methanol containing 1 M HCl for 5 min the concentration of zinc dropped below the EDX detection limit. Similarly, the alumina shell arising from an initial TMA/H_2O cycle can be etched with a weak basic solution [56]. However, conceptually this $TiO_2-Al_2O_3$ core-shell structure may be interesting for DSSC applications [57]. Methyltitanium trichloride or dimethyltitanium dichloride should be reactive enough to replace the initial pulse of TMA or DEZ, but these precursors would need to be synthesized and are unstable [58, 59]. Increasing the deposition temperature to 100 °C lowered the amount of chlorine contaminations. Moreover, using hydrogen peroxide rather than water as oxygen source, $TiO_2\left(\frac{DEZ}{H_2O_2}\middle|\frac{TiCl_4}{H_2O_2}\middle|100\,^\circ C\right)$, resulted in a further improvement of the TiO_2 purity. Regardless, a considerable amount of chlorine impurities remained after calcination ($\approx$1 %).

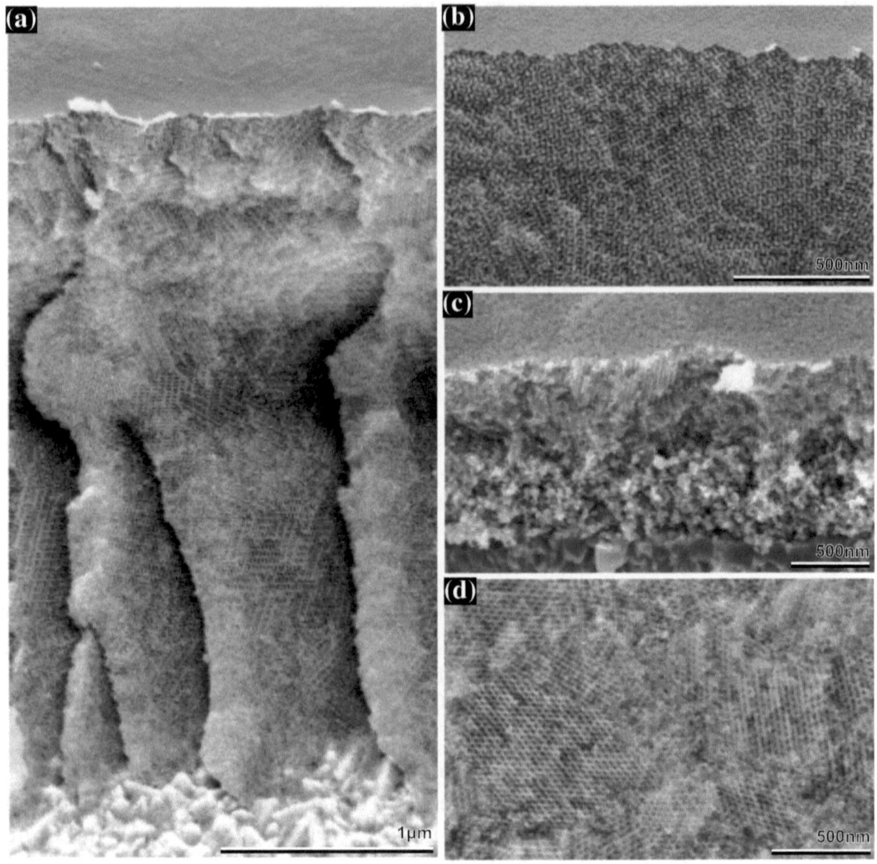

Fig. 8.13 ALD of $ZnO\left(\frac{DEZ}{H_2O}\middle|\frac{DEZ}{H_2O}\middle|100\,°C\right)$ on ozone-treated and thermally annealed templates. **a** Cross-sectional view of the ZnO replica after partial template removal with plasma. **b** Magnified view of the top part of the deposit before RIE. **c** Replica after RIE and subsequent calcination. Note the crumbled layer between the nanostructured part and the substrate. **d** Magnified view of the DG replica after calcination

8.3.5 ALD of Zinc Oxide

We also attempted the refilling of the ozone-treated and thermally annealed templates with $ZnO\left(\frac{DEZ}{H_2O}\middle|\frac{DEZ}{H_2O}\middle|100\,°C\right)$, see Fig. 8.13. DEZ compared to $TiCl_4$ has the advantage that it is more reactive and that the by-product, ethane, has a much lower boiling point than hydrochloric acid, therefore a lower contamination of ligands and by-product is expected. Contrary to TiO_2, the partially structurally unstable part of the replica did not completely vanish during thermal annealing, but rater crumbled and formed a disordered layer between the DG-structured part and the substrate, see Fig. 8.13c.

8.3.6 Dye-Sensitized Solar Cells

$TiO_2(^{TMA}_{H_2O}|^{TiCl_4}_{H_2O}|100\,°C)$, $TiO_2(^{DEZ}_{H_2O}|^{TiCl_4}_{H_2O}|100\,°C)$, and $TiO_2(^{DEZ}_{H_2O_2}|^{TiCl_4}_{H_2O_2}|100\,°C)$ which were deposited on ozone-treated and thermally annealed templates were used to prepare DSSCs after RIE and calcination. The samples were sensitized with the ruthenium-based N-719 dye, capped with a Pt-coated FTO electrode, and finally filled with an iodide/triiodide based liquid ROBUST electrolyte [3, 4]. Although the cells exhibited a relatively high open circuit voltage of $\approx$0.85 V, the overall conversion efficiency was below 0.1 %. We attribute these unsatisfactory performances to the missing crystallinity of the chlorine-doped TiO_2 and the incomplete refilling of the nano-channels. No anatase peaks were observed in XRD spectra and we assume that either confinement effects or the chlorine contamination prevented the crystallization during thermal annealing.

8.4 Concluding Remarks

Even though investing a substantial amount of time and effort into this particular project, we have not quite yet succeeded in preparing DG-structure anatase TiO_2 via the templated ALD for the application in DSSCs. Nevertheless, we evaluate the development of the experimental protocol concerning the substantial enhancement of the thermal stability, solvent resistance, and surface functionality of the mesoporous DG templates as a major success. Very recently we observed, that the thermal stability is further increased to at least 150 °C after exposing the ozone-treated and thermally annealed templates to one cycle of TMA/H_2O. This steep increase in thermal stability can be explained by the subsurface deposition of aluminium oxide containing species forming further interchain cross-linking [18]. These improved material properties might enable application of the chemically modified scaffolds in templating techniques requiring organic solvents which typically would dissolve poly(styrene).

In the near future we will test the newly homebuilt ALD which should facilitate long hold times with a fairly constant precursor concentration. In combination with a deposition temperature of 150 °C and H_2O_2 as oxygen source, we have high hopes to achieve a higher filling fraction than so far and simultaneously overcome the issues concerning the contamination with chlorine.

References

1. A. Yella, H.-W. Lee, H.N. Tsao, C. Yi, A.K. Chandiran, M.K. Nazeeruddin, E.W.-G. Diau, C.-Y. Yeh, S.M. Zakeeruddin, M. Grätzel, Science 334(6056), 629–634 (2011)
2. F. Liu, Y. Liu, L. Huang, X. Hu, B. Dong, W. Shi, Y. Xie, X. Ye, Opt. Commun. **284**(9), 2376–2381 (2011)

3. E.J.W. Crossland, M. Kamperman, M. Nedelcu, C. Ducati, U. Wiesner, D.M. Smilgies, G.E.S. Toombes, M.A. Hillmyer, S. Ludwigs, U. Steiner, H.J. Snaith, Nano Lett. **9**(8), 2807–2812 (2009)
4. E.J.W. Crossland, M. Nedelcu, C. Ducati, S. Ludwigs, M.A. Hillmyer, U. Steiner, H.J. Snaith, Nano Lett. **9**(8), 2813–2819 (2009)
5. F. Endres, S. Zein El Abedin, A.Y. Saad, E.M. Moustafa, N. Borissenko, W.E. Price, G.G. Wallace, D.R. MacFarlane, P.J. Newman, A. Bund, Phys. Chem. Chem. Phys. **10**(16), 2189 (2008)
6. J. Aarik, A. Aidla, H. Mändar, T. Uustare, Appl. Surf. Sci. **172**(1–2), 148–158 (2001)
7. M. Leskelä, M. Ritala, Angew. Chem. Int. Ed. **42**(45), 5548–5554 (2003)
8. R.L. Puurunen, J. Appl. Phys. **97**(12), 121301 (2005)
9. S.M. George, Chem. Rev. **110**(1), 111–131 (2009)
10. M. Ritala, M. Leskela, L. Niinisto, P. Haussalo, Chem. Mater. **5**(8), 1174–1181 (1993)
11. M. Ritala, M. Leskelä, E. Nykänen, P. Soininen, L. Niinistö, Thin Solid Films **225**(1–2), 288–295 (1993)
12. H. Kim, E. Pippel, U.G. ösele, M. Knez. Langmuir **25**(23), 13284–13289 (2009)
13. M. Knez, A. Kadri, C. Wege, U. Gösele, H. Jeske, K. Nielsch, Nano Lett. **6**(6), 1172–1177 (2006)
14. G. Triani, J. Campbell, P. Evans, J. Davis, B. Latella, R. Burford, Thin Solid Films **518**(12), 3182–3189 (2010)
15. J. Aarik, A. Aidla, H. Mändar, V. Sammelselg, J. Cryst. Growth **220**(4), 531–537 (2000)
16. M. Puttaswamy, K.B. Haugshøj, L. Højslet Christensen, P. Kingshott, Chem. Eur. J. **16**(47), 13925–13929 (2010)
17. C.A. Wilson, R.K. Grubbs, S.M. George, Chem. Mater. **17**(23), 5625–5634 (2005)
18. J.C. Spagnola, B. Gong, S.A. Arvidson, J.S. Jur, S.A. Khan, G.N. Parsons, J. Mater. Chem. **20**(20), 4213 (2010)
19. H. Shin, D.-K. Jeong, J. Lee, M.M. Sung, J. Kim, Adv. Mater. **16**(14), 1197–1200 (2004)
20. G. Triani, P.J. Evans, D.J. Attard, K.E. Prince, J. Bartlett, S. Tan, R.P. Burford, J. Mater. Chem. **16**(14), 1355 (2006)
21. M. Knez, K. Nielsch, L. Niinistö, Adv. Mater. **19**(21), 3425–3438 (2007)
22. C. Detavernier, J. Dendooven, S. Pulinthanathu Sree, K.F. Ludwig, J.A. Martens, Chem. Soc. Rev. **40**(11), 5242 (2011)
23. K. Malek, M.-O. Coppens, J. Chem. Phys. **119**(5), 2801–2811 (2003)
24. J.W. Elam, D. Routkevitch, P.P. Mardilovich, S.M. George, Chem. Mater. **15**(18), 3507–3517 (2003)
25. S.K. Karuturi, L. Liu, L.T. Su, Y. Zhao, H.J. Fan, X. Ge, S. He, A.T.I. Yoong, J. Phys. Chem. C **114**(35), 14843–14848 (2010)
26. S.K. Karuturi, L. Liu, L.T. Su, A. Chutinan, N.P. Kherani, T.K. Chan, T. Osipowicz, A.I.Y. Tok, Nanoscale **3**(12), 4951–4954 (2011)
27. R. Gordon, D. Hausmann, E. Kim, J. Shepard, Chem. Vap. Deposition **9**(2), 73–78 (2003)
28. S.O. Kucheyev, J. Biener, T.F. Baumann, Y.M. Wang, A.V. Hamza, Z. Li, D.K. Lee, R.G. Gordon, Langmuir **24**(3), 943–948 (2008)
29. S. Mitchell, A. Poulsson, M. Davidson, R. Bradley, Colloids Surf. B **46**(2), 108–116 (2005)
30. B.W. Callen, J. Vac. Sci. Technol. A **13**(4), 2023 (1995)
31. M.L. Steen, A.C. Jordan, E.R. Fisher, J. Membr. Sci. **204**(1–2), 341–357 (2002)
32. E.S. Brandt, J.M. Grace, J. Vac. Sci. Technol. A 30(1), 01A137 (2012)
33. E. Partouche, D. Waysbort, S. Margel, J. Colloid Interface Sci. **294**(1), 69–78 (2006)
34. H. Kumagai, T. Kusunoki, T. Kobayashi, AZojomo **3**, 1–10 (2007)
35. A.S. Zalusky, R. Olayo-Valles, J.H. Wolf, M.A. Hillmyer, J. Am. Chem. Soc. **124**(43), 12761–12773 (2002)
36. J. Rzayev, M.A. Hillmyer, J. Am. Chem. Soc. **127**(38), 13373–13379 (2005)
37. M.A. Cameron, I.P. Gartland, J.A. Smith, S.F. Diaz, S.M. George, Langmuir **16**(19), 7435–7444 (2000)
38. D.L. Bunbury, Can. J. Chem. **43**(6), 1714–1719 (1965)

39. K. Watanabe, H. Ishikawa, W. Ando, Bull. Chem. Soc. Jpn. **51**(4), 1253–1254 (1978)
40. J. Mack, J.R. Bolton, J. Photochem. Photobiol. A **128**(1–3), 1–13 (1999)
41. N. Jayaswal, S. Sinha, A. Kumar, J. Appl. Polym. Sci. **79**(10), 1735–1748 (2001)
42. D. Vione, V. Maurino, C. Minero, D. Borghesi, M. Lucchiari, E. Pelizzetti, Environ. Sci. Technol. **37**(20), 4635–4641 (2003)
43. A.S.G. Curtis, J. Cell Biol. 1983, 97(5), 1500–1506, PMID: 6355120 PMCID: 2112677.
44. A. Hambardzumyan, S. Biltresse, Y. Dufrêne, J. Marchand-Brynaert, J. Colloid Interface Sci. **252**(2), 443–449 (2002)
45. B. Mailhot, J.L. Gardette, Macromolecules **25**(16), 4119–4126 (1992)
46. D.O.H. Teare, C. Ton-That, R.H. Bradley, Surf. Interface Anal. **29**(4), 276–283 (2000)
47. M. Davidson, S. Mitchell, R. Bradley, Surf. Sci. **581**(2–3), 169–177 (2005)
48. J. López-Gejo, H. Gliemann, T. Schimmel, A.M. Braun, Photochem. Photobiol. **81**(4), 777–782 (2005)
49. M. Tajbakhsh, R. Hosseinzadeh, H. Alinezhad, S. Ghahari, A. Heydari, S. Khaksar, Synthesis **2010**, 490–496 (2011)
50. S. Sato, T. Sakamoto, E. Miyazawa, Y. Kikugawa, Tetrahedron **60**(36), 7899–7906 (2004)
51. K. Biswas, O. Prieto, P.J. Goldsmith, S. Woodward, Angew. Chem. Int. Ed. **44**(15), 2232–2234 (2005)
52. L.A. Hull, I.C. Hisatsune, J. Heicklen, J. Am. Chem. Soc. **94**(14), 4856–4864 (1972)
53. E.G. Derouane, Microporous Mesoporous Mater. **104**(1–3), 46–51 (2007)
54. A. Lucas, I. Derycke, P. Lambin, J.-P. Vigneron, L. Leherte, M. Elanany, J.-M. André, A. Larin, D. Vercauteren, J. Mol. Catal. A: Chem. **305**(1–2), 16–23 (2009)
55. D. Jaguste, S. Bhatia, Chem. Eng. Sci. **50**(2), 167–182 (1995)
56. J.R. Deneault, X. Xiao, T. Kang, J.S. Wang, C.M. Wai, G.J. Brown, M.F. Durstock, ChemPhysChem **13**(1), 256–260 (2012)
57. F. Fabregat-Santiago, J. García-Cañadas, E. Palomares, J.N. Clifford, S.A. Haque, J.R. Durrant, G. Garcia-Belmonte, J. Bisquert, J. Appl. Phys. **96**(11), 6903–6907 (2004)
58. G.L. Karapinka, J.J. Smith, W.L. Carrick, J. Polym. Sci., Part A. Polym. Chem. **50**(153), 143–150 (1961)
59. J. Liu, D. Zhang, J. Huang, Y. Qian, A.S.C. Chan, J. Polym. Sci., Part A. Polym. Chem. **38**(9), 1639–1641 (2000)

Chapter 9
Summary and Suggestions for Further Work

The objective of this thesis was to explore the template-assisted patterning of functional materials on the nanoscale to enhance specific material properties. Self-assembled diblock copolymer scaffolds with the unique double-gyroid morphology were chosen as organic templates. Firstly, a highly reproducible synthesis pathway based on atom transfer radical and organocatalytic ring-opening polymerization in combination with a bifunctional initiator was developed in order to ensure the supply of large quantities of double-gyroid-forming diblock copolymers. Electing non-substituted styrene and lactide as monomers and establishing a rapid thermal annealing protocol made the double-gyroid template preparation more environmentally sustainable and economical, both important requirements when aiming for industrial large-scale applications.

In the second part of the study, different strategies were tested to replicate the templates with inorganic as well as organic functional materials using electrochemical and atomic layer deposition. The different patterned materials included vanadium pentoxide, nickel oxide, platinum, nickel, cobalt, poly(pyrrole), poly(3,4-ethylenedioxythiophene), poly((3-methyl)thiophene), titanium oxide, and zinc oxide. Further, a versatile approach to three-dimensionally nanopatterned metal ceramics via the post-nanostructuring thermal oxidation of the corresponding metals based on the nanoscale Kirkendall effect were presented. The performance of these double-gyroid-structured functional materials were further tested in photovoltaic devices, electrochromic displays, and supercapacitors. Importantly, some of these outperformed state of the art devices demonstrating the decisive effect a three-dimensional nanomorphology, especially the double-gyroid, can have on the properties of functional materials.

Perhaps the most intriguing result of this study, is the versatility of a template-assisted nanostructuring approach based on mesoporous polymer scaffolds with double-gyroid morphology, regarding the numerous replication methods and the various templated functional materials. Another specific application of the double-gyroid morphology could be in a metal-insulator-metal electrostatic capacitor [1]. A possible strategy for such a nanocapacitor is to combine the here studied nanofabrication

M. R. J. Scherer, *Double-Gyroid-Structured Functional Materials*, Springer Theses, DOI: 10.1007/978-3-319-00354-2_9,

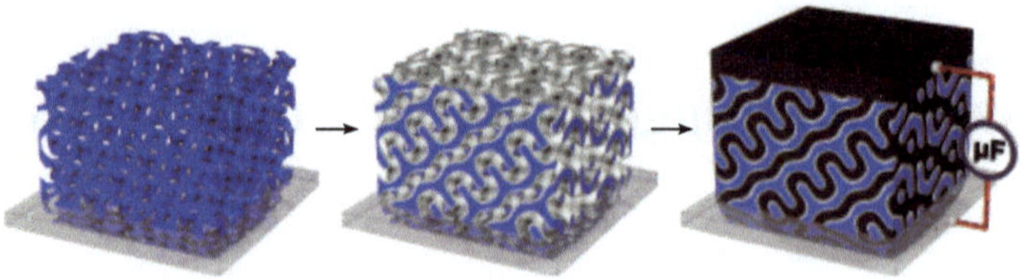

Fig. 9.1 Schematic illustrating the preparation of metal-insulator-metal electrostatic nanocapacitors based on the double-gyroid morphology. Firstly, the DG templates are replicated via nickel electroplating. Subsequent to template dissolution the metal nanostructure is coated with a few nanometer thick electrically insulating alumina layer by ALD. Lastly, a liquified low-melting point alloy is drop-cast onto the sample and allowed to infiltrate the mesopores. The two interdigitated metal networks which are separated by the electrically isolating metal oxide form the active volume V of the electrostatic capacitor

techniques of refilling and coating mesoporous scaffolds, namely templated electroplating of nickel and atomic layer deposition of alumina, with the nanocasting of a low-melting point metal. Figure 9.1 illustrates the three steps required to build such a capacitor. Firstly, a nickel replica of a double-gyroid template is prepared by electroplating (see Chap. 6), secondly after template removal the double-gyroid replica is conformally coated with a few nanometer thick layer of electrically isolating Al_2O_3 using ALD, and finally a low-melting point alloy, such as Field's or Rose's alloy, is liquified and cast into the mesopores. The Field's metal is an eutectic alloy of bismuth (32.5 %), indium (51 %), and tin (16.5 %) which melts at approximately 62 °C [2]. Importantly, the second metal electrode is not prepared by electroless plating, since this wet deposition technique typically requires a non pH-neutral precursor solution that tends to dissolve the thin alumina coating and has issues with complete pore filling [3].

The native oxide layer formed on nickel even at room temperature should promote a fairly conformal nucleation of the ALD alumina. In contrast to the polymeric templates mesoporous nickel exhibits a much higher thermal stability, therefore the ALD step should be relatively straightforward. At this stage of the fabrication process it probably is advisable to test the quality and conformality of the Al_2O_3 coating using cyclic voltammetry. The infiltration of Field's alloy into the alumina coated mesopores might be more challenging due to wetting issues. However, the surface tension of alumina can be altered fairly easily using low molecular weight silanes such as methyltrichlorosilane. Moreover, the alloy's contact angle on copper was found to gradually decrease from 38° to 17° as the reflow temperature increased from 80 to 140 °C [2].

A rough estimate for the capacitance of such a capacitor can be derived by approximating the gyroidal minority network made from nickel to be cylindrical with a strut radius $r_i = 5.5$ nm. For an active capacitor volume V, the cylindrical length L is given by

$$L = \frac{\eta V}{2\pi r_i}, \tag{9.1}$$

where $\eta = 161.4\,\mu m^{-1}$ is the surface area to bulk volume ratio of the double-gyroid formed by P(F)S49-*b*-PLA192$|_{0.33h}^{174°C}$. Considering a 2 nm thick ALD alumina coating, which according to Lin et al. exhibits a breakdown voltage of $17.5\,MV\,cm^{-1}$, the outer radius of the cylindrical capacitor is $r_o = 7.5\,nm$ [4]. The double-gyroid capacitors capacitance per active volume, C/V, is given by

$$\frac{C}{V} = \frac{2\pi k\varepsilon_0 L}{V\ln(r_o/r_i)} \tag{9.2}$$

$$= \frac{k\varepsilon_0\eta}{r_i\ln(r_o/r_i)} \tag{9.3}$$

$$\approx k\varepsilon_0 \cdot 10^{13}\,cm^{-2} \tag{9.4}$$

$$\approx 636\,\mu F\,cm^{-2}\mu m^{-1}, \tag{9.5}$$

where $k = 7.6$ is the dielectric constant of alumina and ε_0 the vacuum permittivity [5]. For a $1\,\mu m$ thick double-gyroid-structured metal-insulator-metal electrostatic nanocapacitor the equivalent planar capacitance is $636\,\mu F\,cm^{-2}$.

References

1. P. Banerjee, I. Perez, L. Henn-Lecordier, S.B. Lee, G.W. Rubloff, Nat. Nanotechnol. **4**(5), 292–296 (2009)
2. E.E.M. Noor, N.M. Sharif, C.K. Yew, T. Ariga, A.B. Ismail, Z. Hussain, J. Alloys Compd. **507**(1), 290–296 (2010)
3. H.-Y. Hsueh, Y.-C. Huang, R.-M. Ho, C.-H. Lai, T. Makida, H. Hasegawa, Adv. Mater. **23**(27), 3041–3046 (2011)
4. H.C. Lin, P.D. Ye, G.D. Wilk, Appl. Phys. Lett. **87**(18), 182904 (2005)
5. M. Groner, J. Elam, F. Fabreguette, S. George, Thin Solid Films **413**(1–2), 186–197 (2002)

Appendix A
NMR Spectra

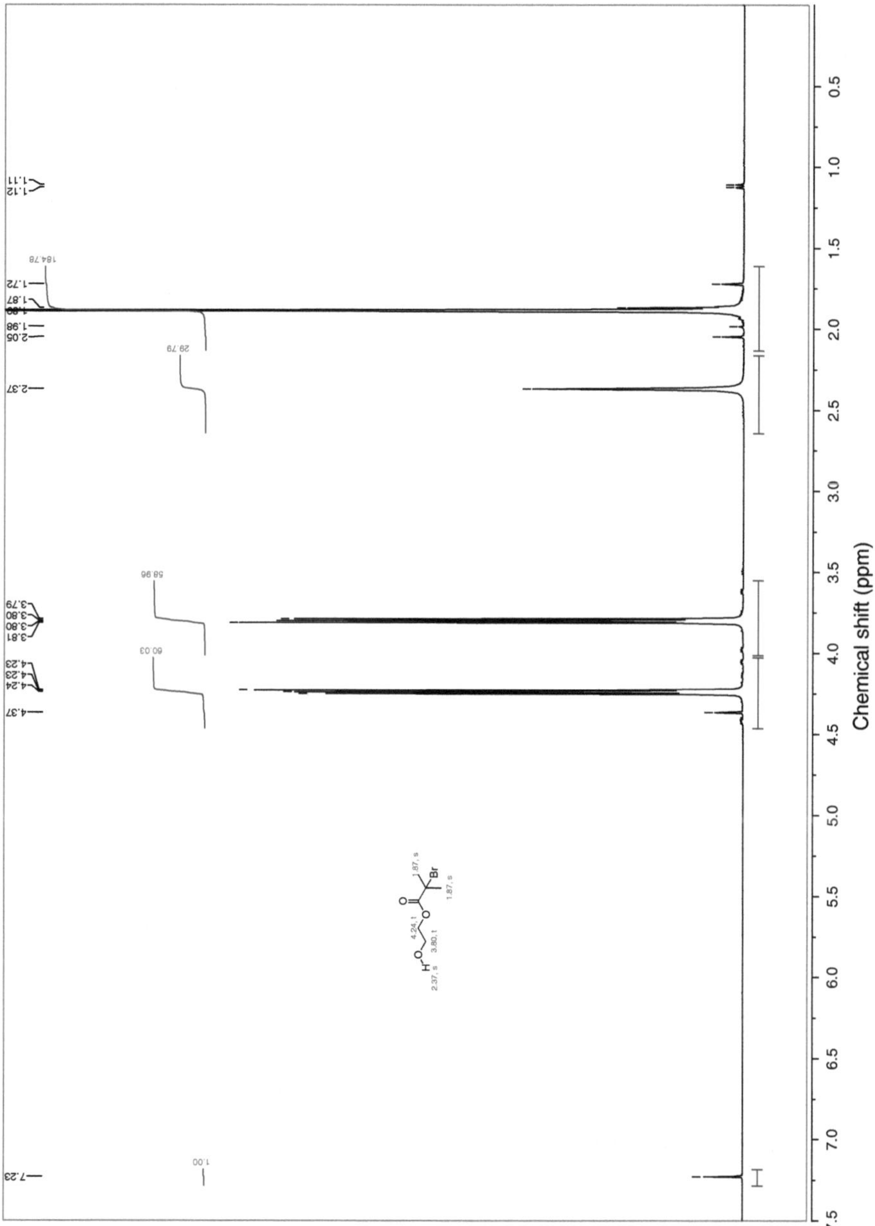

Fig. A.1 ^{1}H NMR (400 MHz, $CDCl_3$) of HEBIB

M. R. J. Scherer, *Double-Gyroid-Structured Functional Materials*,
Springer Theses, DOI: 10.1007/978-3-319-00354-2,

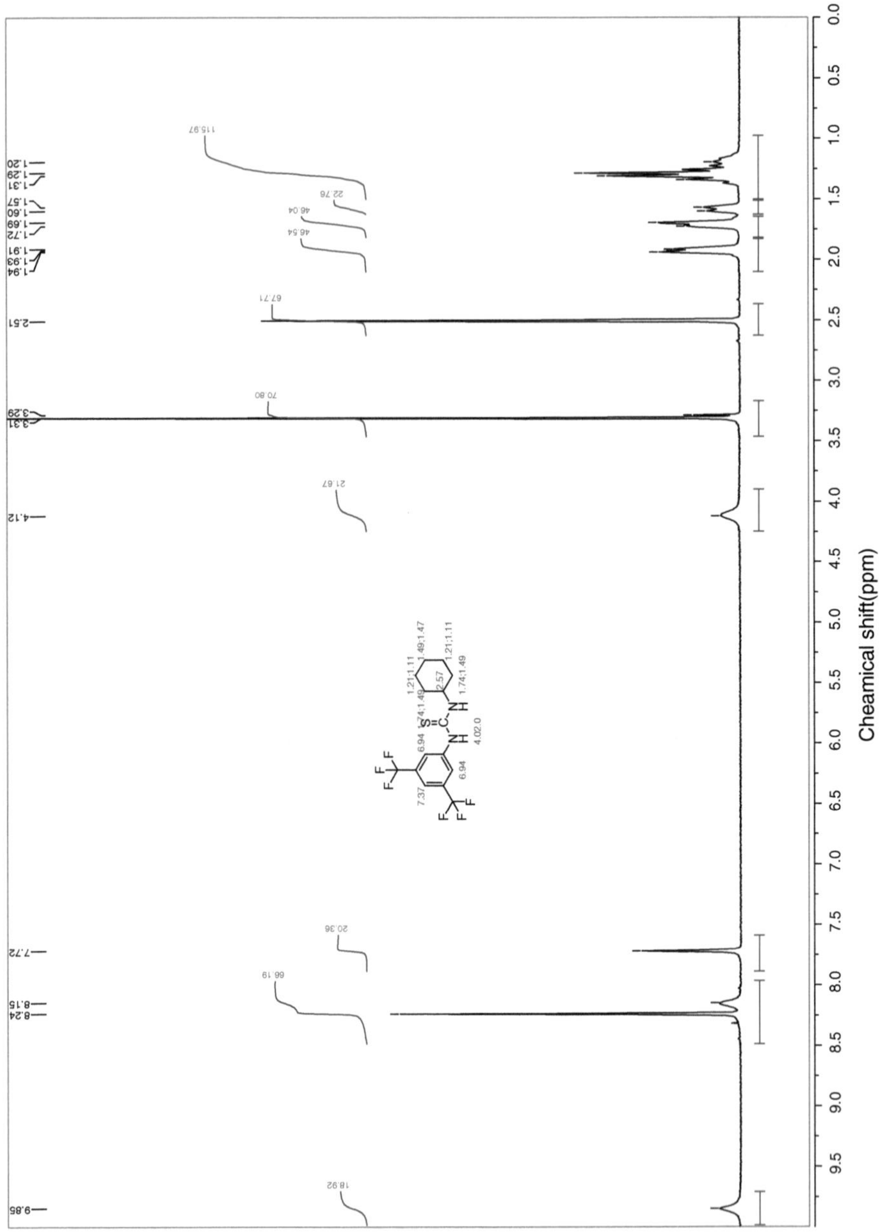

Fig. A.2 ^{1}H NMR (400 MHz, DMSO) of thiourea

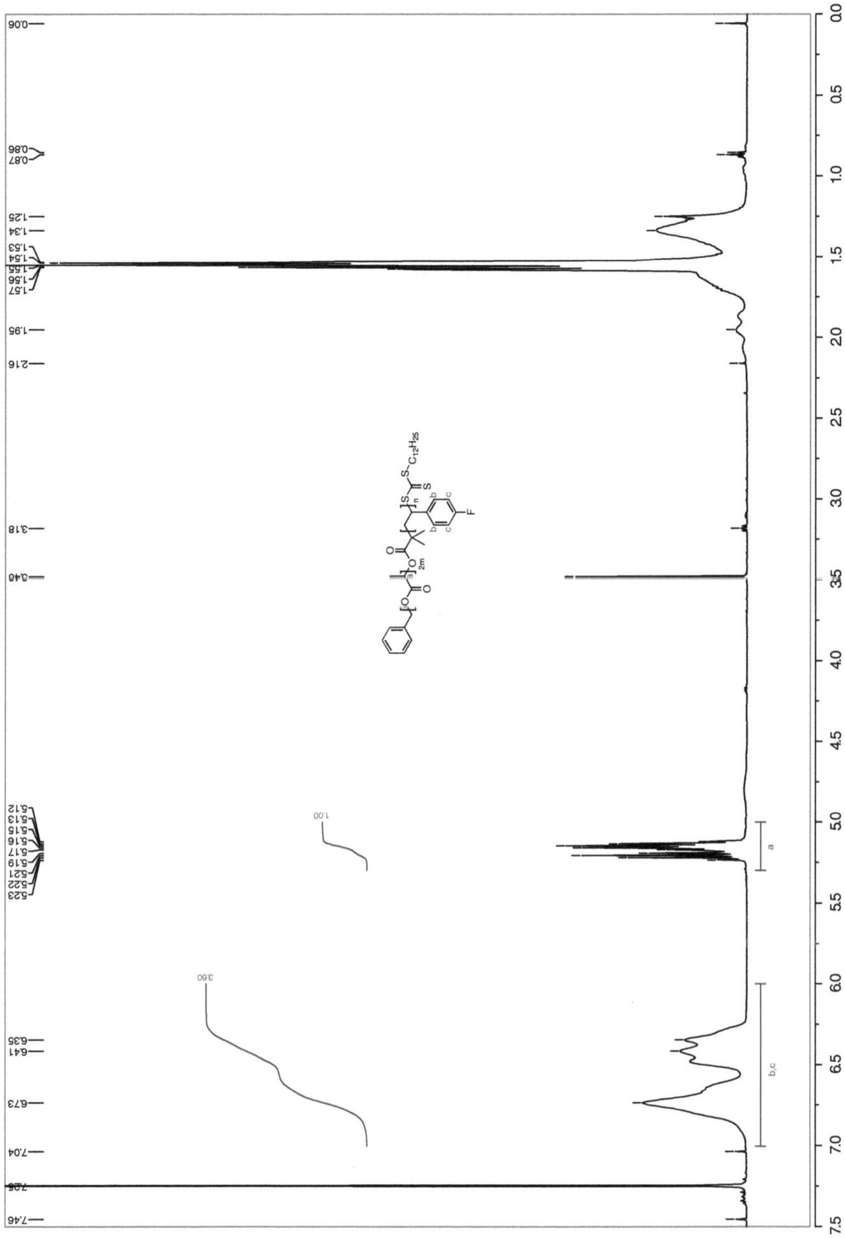

Fig. A.3 ^{1}H NMR (500 MHz, $CDCl_3$) of MHI41

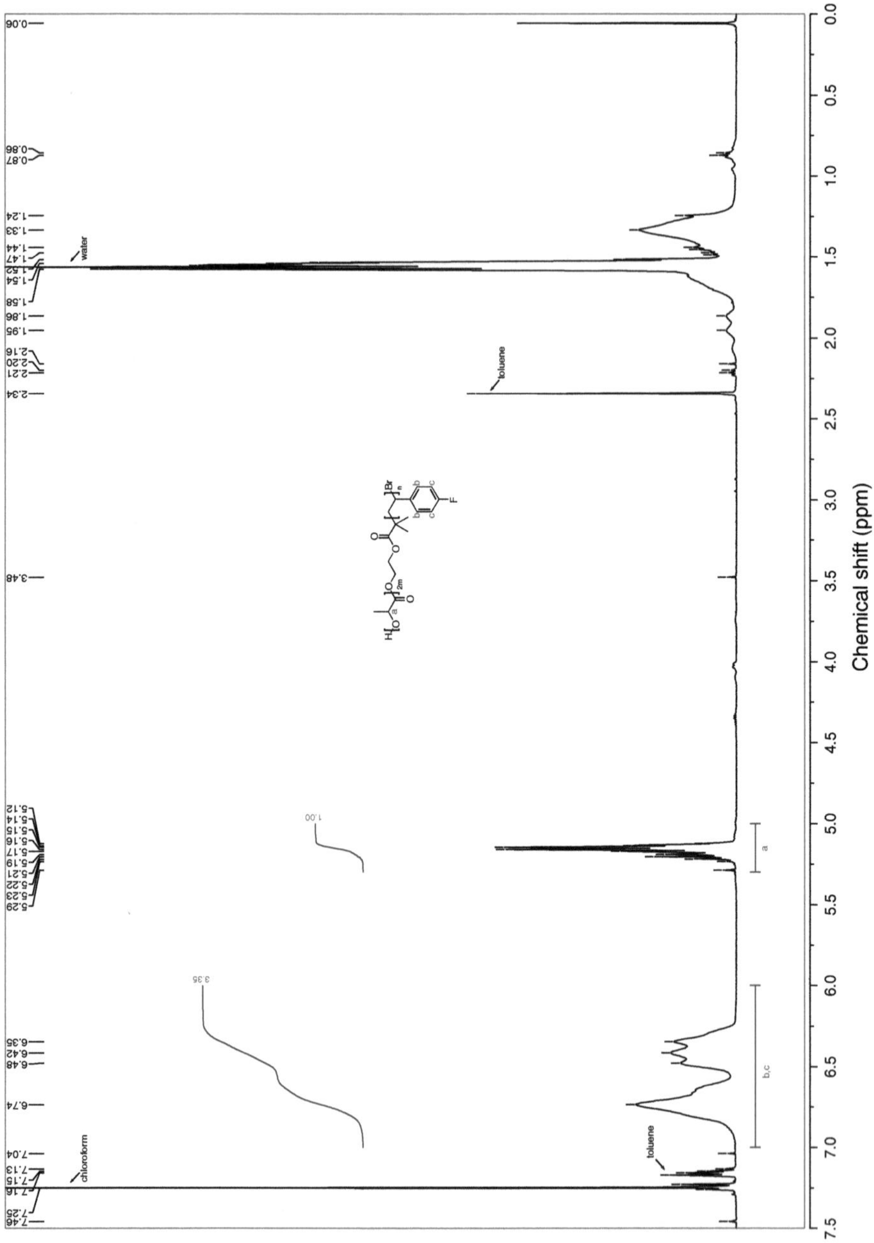

Fig. A.4 ^{1}H NMR (500 MHz, $CDCl_3$) of PFS18-*b*-PLA76

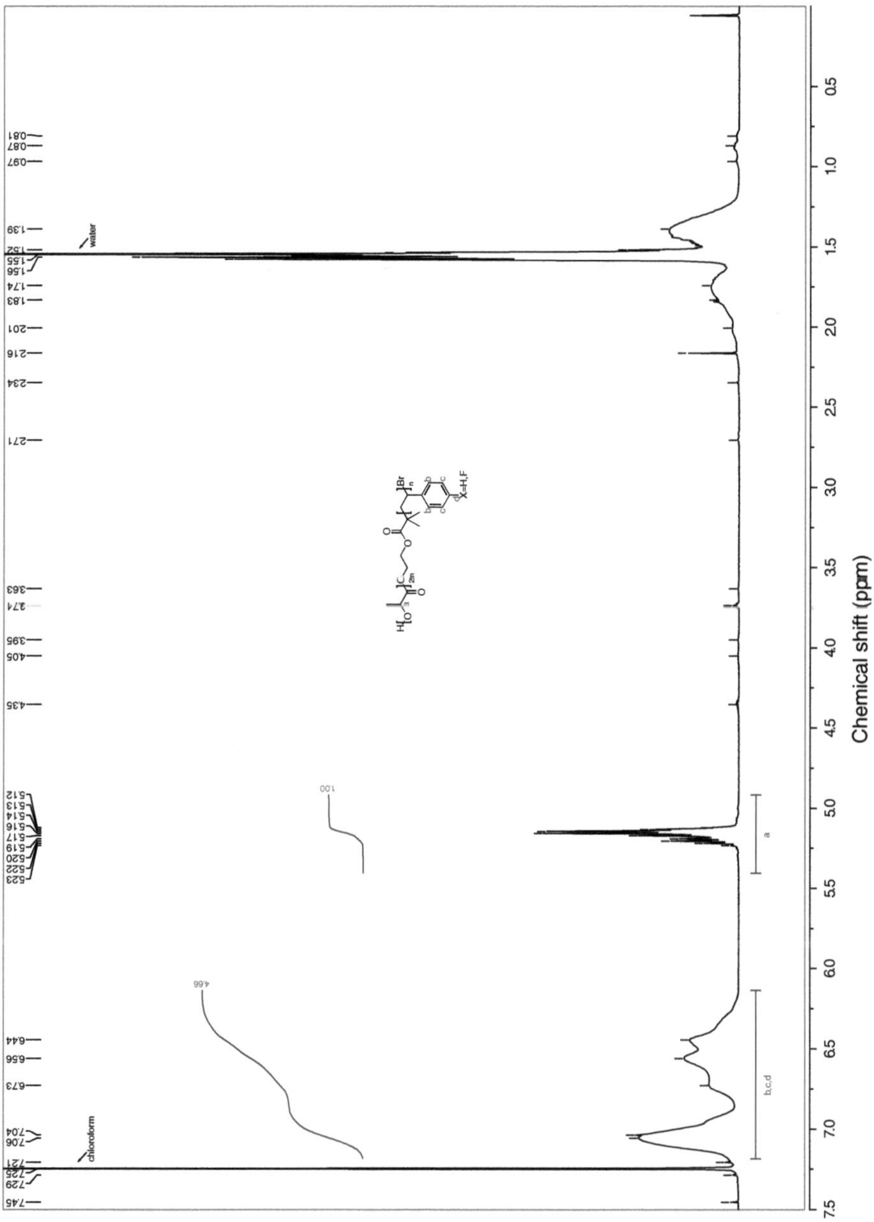

Fig. A.5 ^{1}H NMR (500 MHz, $CDCl_3$) of P(F)S49-*b*-PLA204

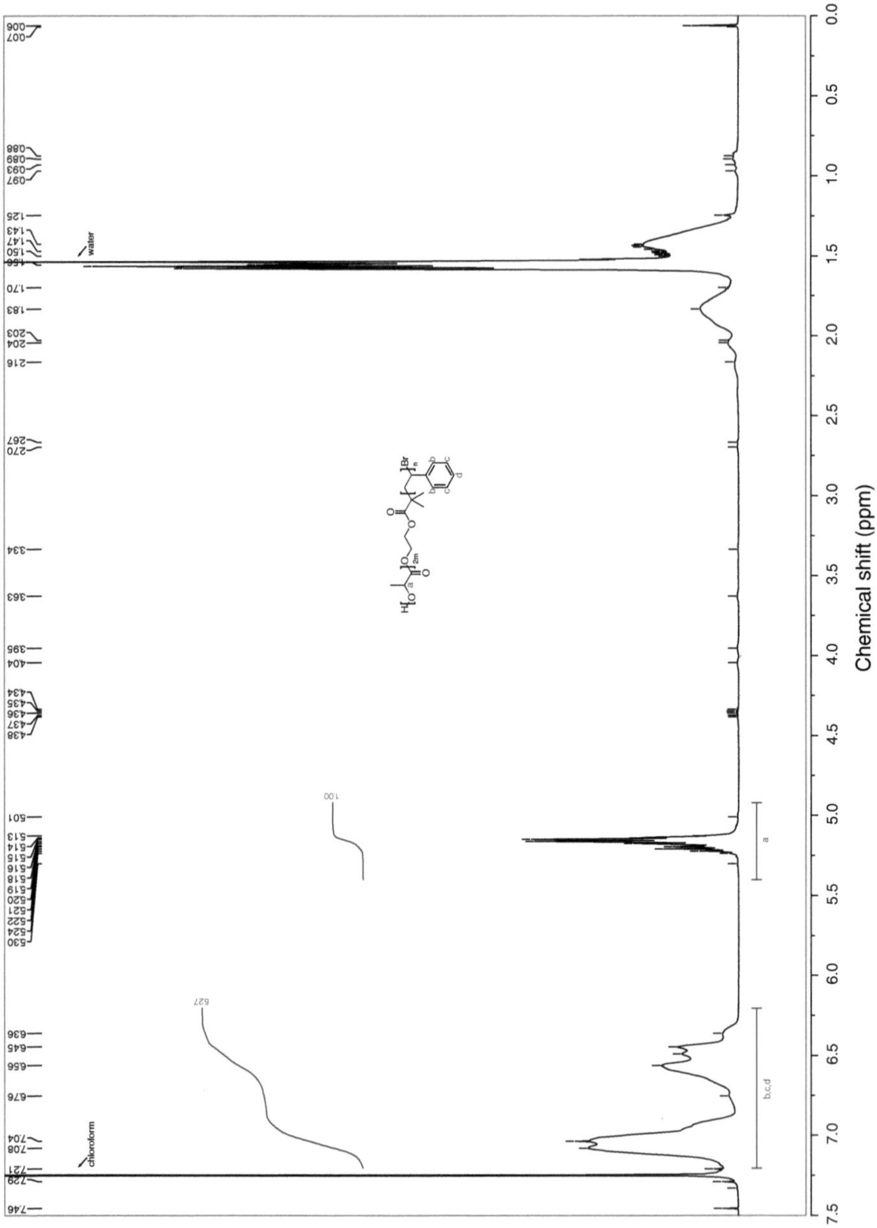

Fig. A.6 ^{1}H NMR (500 MHz, $CDCl_3$) of PS50-*b*-PLA214

Appendix B
List of Diblock Copolymers

Table B.1 Microphase separated structure adopted by PXS-*b*-PLAs for different thermal annealing conditions

Sample	$M_n^{PXS}\left[\frac{kg}{mol}\right]^a$	m_{PXS} [mg][b]	f_{PLA}^{theo} [%][c]	Structure $_t^T$ [°C_h][d]
PFS15-*b*-PLA12	65	25	30.00	X_{19}^{170}
PFS15-*b*-PLA13	65	25	35.00	X_{19}^{170}
PFS15-*b*-PLA14	65	25	40.00	X_{19}^{170}
PFS15-*b*-PLA15	65	25	45.00	X_{19}^{170}
PFS15-*b*-PLA16	65	25	50.00	X_{19}^{170}
PFS17-*b*-PLA17[e]	50	25	40.00	C_{19}^{170}
PFS17-*b*-PLA18[e]	50	25	40.00	C_{19}^{170}
PFS17-*b*-PLA19	50	25	30.00	X_{19}^{170}
PFS17-*b*-PLA20	50	25	35.00	X_{19}^{170}
PFS17-*b*-PLA21	50	25	40.00	X_{19}^{170}
PFS17-*b*-PLA22	50	25	45.00	X_{19}^{170}
PFS17-*b*-PLA23	50	25	50.00	X_{19}^{170}
PFS19-*b*-PLA27[e]	30	50	42.00	W_{19}^{170}
PFS18-*b*-PLA28	15	25	32.50	S_{19}^{170}
PFS18-*b*-PLA29	15	25	35.00	S_{19}^{170}
PFS18-*b*-PLA30	15	25	37.50	S_{19}^{170}
PFS18-*b*-PLA31	15	25	40.00	S_{19}^{170}
PFS18-*b*-PLA32[e]	15	25	42.50	$\mathbf{DG}_{19}^{170}$, W_{-}^{-}
PFS18-*b*-PLA33[e]	15	25	45.00	$^*\mathbf{DG}_{19}^{170}$
PFS19-*b*-PLA34[e]	30	25	32.50	S_{19}^{170}
PFS19-*b*-PLA35	30	25	35.00	S_{19}^{170}
PFS19-*b*-PLA36	30	25	37.50	S_{19}^{170}
PFS19-*b*-PLA37	30	25	40.00	W_{19}^{170}
PFS19-*b*-PLA38	30	25	42.50	W_{19}^{170}
PFS19-*b*-PLA39[e]	30	25	45.00	$[W,L]_{19}^{170}$
PFS17-*b*-PLA40	50	25	32.50	C_{19}^{170}
PFS17-*b*-PLA41	50	25	35.00	X_{19}^{170}
PFS17-*b*-PLA42	50	25	37.50	X_{19}^{170}
PFS17-*b*-PLA43	50	25	40.00	X_{19}^{170}
PFS17-*b*-PLA44	50	25	42.50	X_{19}^{170}
PFS17-*b*-PLA45	50	25	45.00	X_{19}^{170}
PFS18-*b*-PLA46	15	50	39.00	X_{19}^{170}

(continued)

M. R. J. Scherer, *Double-Gyroid-Structured Functional Materials*,
Springer Theses, DOI: 10.1007/978-3-319-00354-2,

Table B.1 (continued)

Sample	$M_n^{PXS}\left[\frac{kg}{mol}\right]^a$	$m_{PXS}[mg]^b$	f_{PLA}^{theo} [%][c]	Structure $_t^T[^{\circ}_{h}C]^d$
PFS18-*b*-PLA47[e]	15	50	41.00	X_{19}^{170}, $[\mathbf{DG},C]_{20}^{190}$, $\mathbf{DG}_{20}^{170}$
PFS18-*b*-PLA48	15	50	43.00	$[X,{}^{\parallel}C]_{19}^{170}$, $[{}^{*}\mathbf{DG},HPL,{}^{\perp}C]_{20}^{165}$
PFS19-*b*-PLA49	30	50	39.00	$[X,W]_{19}^{170}$
PFS19-*b*-PLA50[e]	30	50	41.00	$[X,W]_{19}^{170}$
PFS19-*b*-PLA51	30	50	43.00	W_{19}^{170}
PFS17-*b*-PLA52	50	50	39.00	$[X,W]_{19}^{170}$
PFS17-*b*-PLA53[e]	50	50	41.00	X_{19}^{170}
PFS17-*b*-PLA54	50	50	43.00	W_{19}^{170}
PFS19-*b*-PLA55[e]	30	50	41.00	X_{19}^{170}, $[L,C]_{-}^{-}$
PFS17-*b*-PLA56[e]	50	50	41.00	X_{19}^{170}
PFS19-*b*-PLA57	30	50	37.50	X_{14}^{190}
PFS19-*b*-PLA58	30	50	39.00	X_{14}^{190}, C_{14}^{170}
PFS19-*b*-PLA59	30	50	40.50	X_{14}^{190}
PFS19-*b*-PLA60	30	50	42.00	$[C,L]_{14}^{170}$
PFS19-*b*-PLA61	30	50	43.50	$[C,L]_{14}^{170}$, C_{35}^{180}
PFS19-*b*-PLA62	30	50	45.00	$[C,L]_{14}^{170}$, C_{35}^{180}
PFS17-*b*-PLA63	50	50	37.50	X_{14}^{170}
PFS17-*b*-PLA64	50	50	39.00	W_{14}^{170}
PFS17-*b*-PLA65	50	50	40.50	W_{14}^{170}
PFS17-*b*-PLA66	50	50	42.00	X_{14}^{170}
PFS17-*b*-PLA67	50	50	43.50	X_{14}^{170}
PFS17-*b*-PLA68	50	50	45.00	X_{14}^{170}
PFS18-*b*-PLA75	15	50	41.00	${}^{\perp}C_{20}^{190}$, $[\,\mathbf{DG},{}^{\perp}C]_{17}^{170}$
PFS18-*b*-PLA76[ef]	15	50	43.50	$[{}^{*}\mathbf{DG},HPL,{}^{\perp}C,X]_{17}^{165}$, ${}^{\perp}C_{20}^{190}$, HPL_{39}^{100}, $C_{0.25}^{200}$, X_{17}^{170}
PFS19-*b*-PLA77	30	50	42.50	C_{20}^{190}
PFS19-*b*-PLA78	30	50	45.00	C_{20}^{190}
PFS19-*b*-PLA79[e]	30	50	47.50	C_{20}^{190}
PFS19-*b*-PLA80[e]	30	50	50.00	C_{20}^{190}
PFS18-*b*-PLA81	15	50	41.00	HPL_{-}^{-}, $[{}^{*}\mathbf{DG},HPL,{}^{\perp}C]_{20}^{190}$, ${}^{*}\mathbf{DG}_{2}^{170}$
PFS18-*b*-PLA82[f]	15	50	45.00	HPL_{48}^{150}, $[{}^{*}\mathbf{DG},HPL,{}^{\perp}C,X]_{20}^{190}$
PFS19-*b*-PLA83	30	50	50.00	C_{20}^{190}
PFS19-*b*-PLA84	30	50	53.00	$[C,L]_{20}^{190}$
PFS19-*b*-PLA85	30	50	56.00	$[L,C]_{20}^{190}$
PFS23-*b*-PLA86	24	40	39.00	C_{20}^{190}
PFS23-*b*-PLA87[ef]	24	40	40.50	C_{20}^{190}
PFS23-*b*-PLA88	24	40	42.00	C_{20}^{190}

(continued)

Table B.1 (continued)

Sample	$M_n^{PXS}\left[\frac{kg}{mol}\right]^a$	$m_{PXS}[mg]^b$	f_{PLA}^{theo} [%][c]	Structure $^T_t[^{\circ}_h C]^d$
PFS23-*b*-PLA89	24	40	43.50	C_{20}^{190}
PFS23-*b*-PLA90	24	40	45.00	$[C,HPL]_{20}^{190}$
PFS23-*b*-PLA91[e]	24	40	46.50	$[C,HPL]_{20}^{190}$
PFS23-*b*-PLA92	24	40	48.00	C_{20}^{190}
PFS23-*b*-PLA93	24	40	49.50	C_{20}^{190}
PFS23-*b*-PLA94[ef]	24	40	51.00	L_{30}^{200}
PFS18-*b*-PLA95	15	35	38.75	$^{\parallel}C_{39}^{160}, X_{19}^{175}$
PFS18-*b*-PLA96	15	35	40.00	$\mathbf{DG}_{39}^{160}$
PFS18-*b*-PLA97[f]	15	35	41.25	$\mathbf{DG}_{39}^{160}$
PFS18-*b*-PLA98	15	35	42.50	$\mathbf{DG}_{39}^{160}$
PFS18-*b*-PLA99	15	35	43.75	$^{\parallel}C_{39}^{160}$
PFS18-*b*-PLA100	15	35	45.00	$^{\parallel}C_{39}^{160}, X_{19}^{175}$
PFS23-*b*-PLA101	24	35	37.50	C_{39}^{160}
PFS23-*b*-PLA102	24	35	39.00	C_{39}^{160}
PFS23-*b*-PLA103[ef]	24	35	40.50	C_{39}^{160}
PFS23-*b*-PLA104	24	35	42.00	C_{39}^{160}
PFS23-*b*-PLA105	24	35	43.50	C_{39}^{160}
PFS23-*b*-PLA106	24	35	45.00	C_{39}^{160}
PFS18-*b*-PLA107[g]	15	35	46.25	X_{39}^{160}
PFS18-*b*-PLA108[g]	15	35	47.50	X_{39}^{160}
PFS18-*b*-PLA109[g]	15	35	48.75	X_{39}^{160}
PFS23-*b*-PLA110[g]	24	35	60.00	X_{39}^{160}
PFS23-*b*-PLA111[g]	24	35	65.00	X_{39}^{160}
PFS23-*b*-PLA112[g]	24	35	70.00	X_{39}^{160}
PFS24-*b*-PLA113[g]	21	50	40.00	X_{39}^{160}
PFS24-*b*-PLA114[g]	21	50	42.50	X_{39}^{160}
PFS24-*b*-PLA115[g]	21	50	45.00	X_{39}^{160}
PFS26-*b*-PLA116[ef]	17	500	43.00	C_{24}^{120}
PFS23-*b*-PLA117	24	250	43.00	C_{24}^{120}
PFS26-*b*-PLA118[f]	17	500	38.00	$W_{24}^{120}, {}^{\perp}C_{24}^{190}, \mathbf{DG}_{1}^{170}, \mathbf{DG}_{1}^{160}$
PFS26-*b*-PLA119[f]	17	500	39.00	$C_{24}^{120}, {}^{\perp}C_{24}^{190}, \mathbf{DG}_{0.66}^{180}$
PFS26-*b*-PLA120	17	500	40.00	$C_{24}^{190}, HPL_{1}^{160}, [\mathbf{DG},C]_{0.25}^{210}$
PFS26-*b*-PLA121	17	500	41.00	$C_{24}^{120}, [L,HPL]_{1}^{160}$
PFS26-*b*-PLA122	17	500	42.00	C_{1}^{160}
PFS26-*b*-PLA123	17	500	43.00	$^{\parallel}C_{1}^{200}, [C,HPL]_{1}^{180}$
PFS26-*b*-PLA124	17	500	44.00	$^{\parallel}C_{1}^{200}$

(continued)

Table B.1 (continued)

Sample	$M_n^{PXS}\left[\frac{kg}{mol}\right]^a$	m_{PXS}[mg][b]	f_{PLA}^{theo} [%][c]	Structure $_t^T[^{\circ C}_{h}]^d$
PFS26-*b*-PLA125	17	500	45.00	$^{\parallel}C_{1}^{200}$
PFS26-*b*-PLA126	17	500	37.00	$C_{0.66}^{180}$
PFS26-*b*-PLA127[ef]	17	500	38.00	$\mathbf{DG}_{0.66}^{180}$
PFS27-*b*-PLA128	33	500	37.00	$W_{0.75}^{180}$, $X_{0.75}^{200}$
PFS27-*b*-PLA129	33	500	38.00	$W_{0.75}^{180}$, $X_{0.75}^{200}$
PFS27-*b*-PLA130[f]	33	500	39.00	$W_{0.75}^{180}$, $X_{0.75}^{200}$
PFS27-*b*-PLA131	33	500	40.00	$W_{0.75}^{180}$, $X_{0.75}^{200}$
PFS26-*b*-PLA132[f]	17	500	38.25	$\mathbf{DG}_{0.33}^{190}$, $W_{0.75}^{170}$
PFS26-*b*-PLA133[f]	17	500	38.50	$\mathbf{DG}_{0.33}^{190}$, $W_{0.75}^{170}$
PFS26-*b*-PLA134	17	750	38.00	$\mathbf{DG}_{0.75}^{180}$, $W_{0.75}^{170}$, $\mathbf{DG}_{0.25}^{200}$
PFS26-*b*-PLA135	17	750	38.50	$\mathbf{DG}_{0.75}^{180}$, $[W,\mathbf{DG}]_{0.75}^{170}$
PFS26-*b*-PLA136[e]	17	750	39.00	$\mathbf{DG}_{0.75}^{180}$, $W_{0.75}^{170}$, $\mathbf{DG}_{0.33}^{190}$, $\mathbf{DG}_{0.25}^{200}$
PFS23-*b*-PLA137	24	500	38.00	$C_{0.75}^{180}$, $C_{0.33}^{190}$, $C_{0.25}^{200}$
PFS23-*b*-PLA138	24	500	39.00	$C_{0.75}^{180}$, $C_{0.33}^{190}$, $C_{0.25}^{210}$
PFS24-*b*-PLA139	21	500	38.00	$C_{0.75}^{180}$, $C_{0.33}^{190}$, $C_{0.25}^{210}$
PFS24-*b*-PLA140	21	500	39.00	$C_{0.75}^{180}$, $HPL_{0.33}^{190}$
PFS24-*b*-PLA141	21	500	37.00	$C_{0.33}^{190}$
PFS24-*b*-PLA142	21	500	40.00	$C_{0.33}^{190}$
PFS24-*b*-PLA143	21	500	41.40	$C_{0.33}^{190}$
PFS18-*b*-PLA144	15	500	38.50	$\mathbf{DG}_{0.33}^{190}$, $\mathbf{DG}_{0.33}^{183}$
PFS35-*b*-PLA146	–	500	38.50	$[L,C]_{0.33}^{183}$
PFS36-*b*-PLA147	–	500	38.50	$L_{0.33}^{183}$
PFS37-*b*-PLA148[ef]	20	500	38.50	$W_{0.33}^{183}$, $C_{0.16}^{195}$, $C_{0.1}^{210}$
PFS38-*b*-PLA149	19	500	38.50	$W_{0.33}^{183}$, $C_{0.16}^{195}$, $C_{0.1}^{210}$, $C_{0.05}^{235}$
P(F)S39-*b*-PLA150[f]	19	500	38.50	$W_{0.33}^{183}$, $W_{0.16}^{195}$, $\mathbf{DG}_{0.1}^{210}$
P(F)S40-*b*-PLA151	17	500	38.50	$W_{0.33}^{183}$, $\mathbf{DG}_{0.16}^{195}$, $\mathbf{DG}_{0.1}^{210}$, $\mathbf{DG}_{0.16}^{195}$
P(F)S41-*b*-PLA152	18	500	38.50	$W_{0.33}^{183}$, $^{\parallel}C_{0.16}^{195}$
PFS26-*b*-PLA118	17	500	38.00	$\mathbf{DG}_{0.33}^{183}$, $\mathbf{DG}_{0.25}^{175}$
PFS37-*b*-PLA153	20	500	38.00	$^{\parallel}C_{0.16}^{195}$, $C_{0.16}^{181}$
PFS38-*b*-PLA154	19	500	38.00	$^{\parallel}C_{0.16}^{195}$, $C_{0.16}^{181}$, $C_{0.1}^{210}$, $X_{0.05}^{240}$
P(F)S39-*b*-PLA155	19	500	38.00	$W_{0.16}^{181}$, $\mathbf{DG}_{0.1}^{210}$
P(F)S40-*b*-PLA156	17	500	38.00	$\mathbf{DG}_{0.16}^{181}$, $\mathbf{DG}_{0.1}^{210}$
P(F)S41-*b*-PLA157	18	500	38.00	$W_{0.16}^{181}$, $\mathbf{DG}_{0.1}^{210}$
PFS37-*b*-PLA158	20	500	37.50	$^{\parallel}C_{0.16}^{195}$, $^{\parallel}C_{0.1}^{210}$
PFS38-*b*-PLA159	19	500	37.50	$^{\parallel}C_{0.16}^{195}$, $^{\parallel}C_{0.1}^{210}$, $X_{0.05}^{240}$
P(F)S39-*b*-PLA160	19	500	37.50	$\mathbf{DG}_{0.16}^{195}$, $\mathbf{DG}_{0.1}^{210}$

(continued)

Table B.1 (continued)

Sample	$M_n^{PXS}\left[\frac{kg}{mol}\right]^a$	$m_{PXS}[mg]^b$	f_{PLA}^{theo} [%][c]	Structure $^{T}_{t}[^{\circ C}_{h}]^d$
P(F)S40-*b*-PLA161	17	500	37.50	$\mathbf{DG}^{210}_{0.1}$
P(F)S41-*b*-PLA162	18	500	37.50	$\mathbf{DG}^{210}_{0.1}$
PFS18-*b*-PLA163	15	500	38.30	$\mathbf{DG}^{181}_{0.33}$, $^{\parallel}C^{205}_{0.16}$
PFS38-*b*-PLA164	19	500	39.00	$C^{181}_{0.33}$, $C^{205}_{0.16}$, $X^{235}_{0.05}$
P(F)S39-*b*-PLA165	19	1000	38.00	$W^{181}_{0.33}$, $[\mathbf{DG},C]^{205}_{0.16}$, $C^{235}_{0.05}$
P(F)S40-*b*-PLA166	17	1000	38.00	$W^{181}_{0.33}$, $\mathbf{DG}^{205}_{0.16}$, $X^{235}_{0.05}$
P(F)S41-*b*-PLA167	18	1000	38.00	$W^{181}_{0.33}$, $\mathbf{DG}^{205}_{0.16}$, $X^{235}_{0.05}$, $X^{205}_{0.55}$
P(F)S42-*b*-PLA168	17	500	38.00	$W^{181}_{0.33}$, $\mathbf{DG}^{205}_{0.16}$, $X^{235}_{0.05}$, $X^{205}_{0.55}$
P(F)S42-*b*-PLA169	17	500	37.50	$[\mathbf{DG},W]^{181}_{0.33}$, $\mathbf{DG}^{205}_{0.16}$, $X^{235}_{0.05}$, $X^{205}_{0.55}$
PFS43-*b*-PLA170	19	500	38.00	$C^{181}_{0.16}$, $C^{210}_{0.1}$
P(F)S42-*b*-PLA171	17	500	37.00	$^{\parallel}C^{180}_{0.25}$, $^{\parallel}C^{205}_{0.16}$
PFS44-*b*-PLA172	15	500	38.00	$\mathbf{DG}^{180}_{0.25}$, $^{\parallel}C^{205}_{0.16}$, $W^{170}_{0.25}$
PFS44-*b*-PLA173	15	500	38.50	$\mathbf{DG}^{180}_{0.25}$, $[\mathbf{DG},C]^{205}_{0.16}$, $W^{170}_{0.25}$
PFS45-*b*-PLA174	15	500	38.00	$\mathbf{DG}^{180}_{0.25}$, $^{\perp}C^{205}_{0.16}$, $\mathbf{DG}^{170}_{0.25}$
PFS45-*b*-PLA175	15	500	38.50	$\mathbf{DG}^{180}_{0.25}$, $^{\perp}C^{205}_{0.16}$
P(F)S46-*b*-PLA176	18	500	37.00	$W^{180}_{0.25}$, $C^{205}_{0.16}$
P(F)S46-*b*-PLA177	18	500	38.00	$W^{180}_{0.25}$, $\mathbf{DG}^{205}_{0.16}$
PS47-*b*-PLA178	17	500	36.00	$W^{180}_{0.25}$, $^{\parallel}C^{205}_{0.16}$
PS47-*b*-PLA179	17	500	37.00	$W^{180}_{0.25}$, $^{\parallel}C^{205}_{0.16}$
PS47-*b*-PLA180	17	500	38.00	$W^{180}_{0.25}$, $C^{205}_{0.16}$
PFS38-*b*-PLA181	19	500	40.00	$[HPL,C]^{180}_{0.25}$, $C^{205}_{0.16}$
P(F)S42-*b*-PLA182	17	500	38.50	$C^{180}_{0.25}$, $C^{205}_{0.16}$
PFS44-*b*-PLA183	15	500	39.00	$[HPL,\mathbf{DG}]^{180}_{0.25}$, $[\mathbf{DG},C]^{205}_{0.16}$
PFS45-*b*-PLA184	15	1500	38.25	$\mathbf{DG}^{180}_{0.16}$, $\mathbf{DG}^{180}_{0.25}$, $^{\perp}C^{205}_{0.16}$
P(F)S46-*b*-PLA185	18	500	38.50	$W^{180}_{0.25}$, $[W,\mathbf{DG}]^{205}_{0.16}$
P(F)S46-*b*-PLA186	18	500	39.00	$W^{180}_{0.25}$, $W^{205}_{0.16}$
PS47-*b*-PLA187	17	500	38.50	$W^{180}_{0.25}$, $W^{205}_{0.16}$
PS47-*b*-PLA188	17	500	39.00	$W^{180}_{0.25}$, $W^{205}_{0.16}$
PS47-*b*-PLA189	17	500	39.50	$W^{180}_{0.25}$, $W^{205}_{0.16}$
PFS45-*b*-PLA190[g]	15	600	38.25	$X^{180}_{0.25}$, $X^{205}_{0.16}$
PFS48-*b*-PLA191[g]	11	500	38.25	$X^{180}_{0.25}$, $X^{205}_{0.16}$
P(F)S49-*b*-PLA192[e]	14	500	38.25	$\mathbf{DG}^{180}_{0.25}$, $^{\parallel}C^{205}_{0.16}$, $\mathbf{DG}^{175}_{0.25}$
PS50-*b*-PLA193[g]	13	500	38.25	$X^{180}_{0.25}$, $X^{205}_{0.16}$
PFS35-*b*-PLA194	–	500	38.00	$C^{180}_{0.25}$, $C^{205}_{0.16}$
PFS36-*b*-PLA195	–	500	38.00	$L^{180}_{0.25}$, $X^{205}_{0.16}$
PFS34-*b*-PLA196	–	500	38.25	$W^{180}_{0.25}$, $W^{205}_{0.16}$
PFS48-*b*-PLA198	11	100	38.25	$X^{180}_{0.25}$, $X^{205}_{0.16}$
PS50-*b*-PLA199	13	500	38.25	$^{\parallel}C^{180}_{0.25}$, $^{\parallel}C^{205}_{0.16}$

(continued)

Table B.1 (continued)

Sample	$M_n^{PXS}\left[\frac{kg}{mol}\right]^a$	m_{PXS}[mg][b]	f_{PLA}^{theo} [%][c]	Structure $_t^T[^{\circ C}_{h}]^d$
PS50-*b*-PLA201	13	500	37.25	$\mathbf{DG}^{180}_{0.25}$, $^{\perp}C^{205}_{0.16}$
PS50-*b*-PLA202	13	500	39.25	$\mathbf{DG}^{180}_{0.25}$, $^{\perp}C^{205}_{0.16}$
PS50-*b*-PLA203	13	500	40.25	$[\mathbf{DG},HPL]^{180}_{0.25}$, $^{\perp}C^{205}_{0.16}$
P(F)S49-*b*-PLA204[ef]	14	500	38.25	$\mathbf{DG}^{180}_{0.25}$
P(F)S49-*b*-PLA205	14	500	38.25	$\mathbf{DG}^{180}_{0.25}$
PS50-*b*-PLA206	13	500	37.60	$\mathbf{DG}^{180}_{0.25}$
P(F)S41-*b*-PLA207	18	500	39.30	$W^{174}_{0.33}$, $\mathbf{DG}^{210}_{0.1}$
PS47-*b*-PLA208	17	500	37.30	$W^{174}_{0.33}$
P(F)S49-*b*-PLA209	14	1000	38.25	$\mathbf{DG}^{174}_{0.33}$
P(F)S49-*b*-PLA210	14	1000	38.25	$\mathbf{DG}^{174}_{0.33}$
P(F)S49-*b*-PLA211	14	1000	38.25	$\mathbf{DG}^{174}_{0.33}$
P(F)S49-*b*-PLA212	14	1000	38.25	$\mathbf{DG}^{174}_{0.33}$
P(F)S49-*b*-PLA213	14	1000	38.25	$C^{174}_{0.33}$
PS50-*b*-PLA214[e]	13	1000	37.50	$\mathbf{DG}^{174}_{0.33}$
PS50-*b*-PLA215	13	1000	38.00	$\mathbf{DG}^{174}_{0.33}$
PFS26-*b*-PLA120:126 = 1:1			38.50	$C^{180}_{0.75}$, $\mathbf{DG}^{170}_{0.25}$
PFS26-*b*-PLA120:PFS29 = 19:1			38.50	$C^{180}_{0.75}$, $\mathbf{DG}^{170}_{0.25}$
PBrS3-*b*-PLA1	11	500	37.00	$W^{165}_{0.33}$, $C^{180}_{0.25}$, $C^{205}_{0.16}$
PBrS3-*b*-PLA2	11	500	38.00	$W^{165}_{0.33}$, $^{\parallel}C^{180}_{0.25}$, $^{\parallel}C^{205}_{0.16}$
PBrS3-*b*-PLA3	11	500	39.00	$W^{165}_{0.33}$, $^{\parallel}C^{180}_{0.25}$, $^{\parallel}C^{205}_{0.16}$
PBrS3-*b*-PLA4	11	500	46.70	$L^{180}_{0.25}$, $L^{205}_{0.16}$
PClS3-*b*-PLA1	11	500	36.00	$W^{165}_{0.33}$, $^{\parallel}C^{180}_{0.25}$, $^{\parallel}C^{205}_{0.16}$
PClS3-*b*-PLA2	11	500	37.00	$W^{165}_{0.33}$, $^{\parallel}C^{180}_{0.25}$, $^{\parallel}C^{205}_{0.16}$
PClS3-*b*-PLA3	11	500	38.00	$W^{165}_{0.33}$, $^{\parallel}C^{180}_{0.25}$, $^{\parallel}C^{205}_{0.16}$
PClS3-*b*-PLA4	11	500	39.00	$W^{165}_{0.33}$, $^{\parallel}C^{180}_{0.25}$, $^{\parallel}C^{205}_{0.16}$
PClS3-*b*-PLA5	11	500	35.00	$S^{165}_{0.33}$, $S^{180}_{0.25}$, $S^{205}_{0.16}$
PClS3-*b*-PLA6	11	500	40.00	$W^{165}_{0.33}$, $^{\parallel}C^{180}_{0.25}$, $^{\parallel}C^{205}_{0.16}$
PClS3-*b*-PLA7	11	500	41.00	$[C,L]^{180}_{0.25}$, $^{\parallel}C^{205}_{0.16}$
PClS3-*b*-PLA8	11	500	42.00	$[C,L]^{180}_{0.25}$, $^{\parallel}C^{205}_{0.16}$
PClS3-*b*-PLA9	11	500	43.00	$[C,L]^{180}_{0.25}$, $^{\parallel}C^{205}_{0.16}$

[a]Molecular weight of PXS-OH determined by SEC calibrated withpoly(styrene) standards.
[b]Amount of macroinitiator PXS-OH used.
[c]Theoretical volume fraction of PLA assuming $\rho_{PLA} = 1.18\,g\,cm^{-3}$.
[d]Morphologies observed in microphase separated films after thermal annealing: No structure, X; spheres, S; cylinders, C; cylinders orientated parallel to the substrate, $^{\parallel}$C; cylinders orientated perpendicular to the substrate, $^{\perp}$C; bicontinuous worm-like structure, W; lamellae, L; hexagonally perforated lamellae, HPL; double-gyroid, **DG**; double-gyroid crystals, ***DG**. Coexisting morphologies are given in square brackets. The annealing temperature [°C] is given as superscript and the annealing time [h] as subscript.
[e]For SEC data see Table 3.6.
[f]For NMR data see Table 3.6.
[g]PLA synthesis not quenched with benzoic acid.

Appendix C
Curriculum Vitae

Personal details

Name:	Maik Rudolf Johann Scherer
Date of birth:	26th January 1982
Birth place:	Otjiwarongo (Namibia)
Nationality:	German

Education

04/2009–08/2012	**PhD in Physics**, University of Cambridge (UK) Thesis: Double-Gyroid-Structured Functional Materials
10/2005–04/2009	**Diploma in Economics**, University of Konstanz (Germany) Thesis: Decision-theoretic Models of Self-control Final grade: 1.79 (good)
10/2003–01/2009	**Diploma in Physics**, University of Konstanz (Germany) and University of Cambridge (UK) Thesis: Nanostructured Materials via Self-Assembled Templates Final grade: 1.12 (very good)
09/1994–06/2002	A-levels, Werdenfels-Gymnasium in Garmisch-Partenkirchen (Germany) Final grade: 1.3 (very good)
01/1989–08/1994	German School Otjiwarongo (Namibia)

Employment and work experience

since 08/2012	Postdoctoral researcher, Cavendish Laboratory/Nanoscience Centre, University of Cambridge (UK)
04/2009–08/2012	Supervision of students and laboratory training courses at the Department of Physics and NanoDTC program, University of Cambridge (UK)
08/2004–09/2004	Employee at the R&D department of LIC® Langmatz, Garmisch-Partenkirchen (Germany): Programming of measuring and control software
05/2003–09/2003	Internship at the R&D department of LIC® Langmatz, Garmisch-Partenkirchen (Germany): Design and implementation of electronic circuits
1998–2003	Writing computer programs for local IT companies and tutoring high school pupils

Military service

07/2002–03/2003	Gebirgsstabs-Fernmeldelehrbataillon 8, Mittenwald (Germany)

Scholarships

04/2009–08/2012	Nokia Scholarship (4-year PhD funding)
12/2008–04/2009	Dr. Jost-Henkel Scholarship

Publications

Nano Lett. 2013	Efficient Electrochromic Devices Made from 3D Nanotubular Gyroid Networks, M. R. J. Scherer, U. Steiner, accepted.
Nano Lett. 2012	A nanostructured electrochromic supercapacitor, D. Wei♣, M. R. J. Scherer♣, C. Bower, P. Andrew, T. Ryhänen, U. Steiner, *12*(4), 1857–1862, ♣these authors contributed equally.

M. R. J. Scherer, *Double-Gyroid-Structured Functional Materials*,
Springer Theses, DOI: 10.1007/978-3-319-00354-2,

Adv. Mater. 2012	Enhanced electrochromism in Gyroid-Structured vanadium pentoxide, M. R. J. Scherer, L. Li, P. M. S. Cunha, O. A. Scherman, U. Steiner, *24*(9), 1217–1221.
Nat. Nano. 2010	Mimicking the colourful wing scale structure of the papilio blumei butterfly, M. Kolle, P. M. Salgard-Cunha, M. R. J. Scherer, F. Huang, P. Vukusic, S. Mahajan, J. J. Baumberg, U. Steiner, *5*(7), 511–515.
Adv. Mater. 2009	Nanostructured calcite single crystals with gyroid morphologies, A. S. Finnemore, M. R. J. Scherer, R. Langford, S. Mahajan, S. Ludwigs, F. C. Meldrum, U. Steiner, *21*(38–39), 3928–3932.

MIX
Papier aus verantwortungsvollen Quellen
Paper from responsible sources
FSC® C105338

If you have any concerns about our products,
you can contact us on
ProductSafety@springernature.com

In case Publisher is established outside the EU,
the EU authorized representative is:
Springer Nature Customer Service Center GmbH
Europaplatz 3, 69115 Heidelberg, Germany

Printed by Libri Plureos GmbH
in Hamburg, Germany